The International Astronomical Union

Johannes Andersen • David Baneke •
Claus Madsen

The International Astronomical Union

Uniting the Community for 100 Years

Johannes Andersen
Dark Cosmology Centre
The Niels Bohr Institute
University of Copenhagen
Copenhagen, Denmark

Stellar Astrophysics Centre, Department of
Physics and Astronomy
Aarhus University
Aarhus, Denmark

Claus Madsen
Institute of Physics and Astronomy
Aarhus University
Aarhus, Denmark

David Baneke
History and Philosophy of Science
Utrecht University
Utrecht, The Netherlands

ISBN 978-3-319-96964-0   ISBN 978-3-319-96965-7   (eBook)
https://doi.org/10.1007/978-3-319-96965-7

Library of Congress Control Number: 2018965606

Cover illustration: IAU members voting on the Pluto / Dwarf Planet resolution at the IAU General Assembly
in Prague 2006.
Credit: IAU/Robert Hurt (SSC)

This Springer imprint is published by the registered company Springer Nature Switzerland AG
The registered company address is: Gewerbestrasse 11, 6330 Cham, Switzerland

Foreword

Throughout the ages, humanity has been fascinated by the night sky. Questions concerning the nature and evolution of the Universe appeal to a deep human desire to understand our origins and place in the Universe, as we begin to learn that our Solar System—and probably life itself—is not unique.

In search for answers to questions such as these, Western astronomers came together in 1919 in Brussels to found the International Astronomical Union (IAU). A century later, both the science and the world have changed beyond recognition. By combining observations from the entire electromagnetic spectrum and of particles and very recently also gravitational waves, the Universe has revealed itself to be more diverse and dynamic than anyone then imagined. All these discoveries have been made possible by crucial technological advancements and international collaboration. The IAU itself has become fully global, with members from more than 100 countries, but its mission remains the same: to promote and safeguard astronomy in all its aspects through international cooperation. On the way, it has evolved from a Union that was primarily concerned with astronomical research to one that is engaged in society: educational, outreach, and development programmes have been added to its portfolio.

The IAU's 100th anniversary provides an opportunity to highlight the science, technology, and inspiration that astronomy has brought us, with the IAU involved in making all this happen. Today, our emphasis is on making our face to the world more explicit, showing that astronomy is an exciting topic that is not just important for astronomy's own sake, but is important for stimulating young people's interest in the sciences in general. We also realise the growing need for a multidisciplinary approach to some of the questions that we astronomers ask.

This book highlights the major milestones in the IAU history in this century, culminating in its efforts to bring the worldwide community together while also opening our science to the wider public. Unlike Blaauw's 1994 archival history of the IAU, it is written on the combined backgrounds in contemporary astronomy, history of astronomy and science in general, communication, science policy, and insight in the IAU machinery. Numerous interviews with living IAU Presidents, General Secretaries, and others bring the role of the IAU in advancing astronomy worldwide alive. The book thus shows the gradual transformation of the IAU from a Union that was focused exclusively on astronomical research to one that is firmly engaged with society. Enjoy this fascinating book!

President of the IAU, Paris, France   Ewine van Dishoeck
October 2018

Acknowledgements

The book is based on a wide range of sources. It is mainly based on the IAU archives and the Transactions, Information Bulletins, GA Newspapers, etc. But we have also gratefully used the work of many astronomers and historians who went before us, as is illustrated by the bibliography, although many more authors contributed indirectly by providing general background or provoking new thoughts. We should especially mention the Centenary Symposium *Under One Sky* (IAU Symposium 349), held in Vienna in 2018. Its proceedings will be published at the same time as this book.

Finally, we would like to thank:

— All the interviewees who opened their homes and hearts and spoke with great honesty and enthusiasm about their personal histories.
— The IAU officers: IAU Presidents Norio Kaifu, Silvia Torres-Peimbert, and Ewine van Dishoeck and General Secretaries Thierry Montmerle and Piero Benvenuti, who gave our project full support.
— Maria Rosaria d'Antonio and Madeleine Smith-Spanier for welcoming David so warmly during his visits to the archives at the IAU office.
— Ginette Rude and Adriaan Blaauw for making the IAU archives so easy to navigate.
— David DeVorkin and Ian Corbett for many valuable comments and corrections (but any remaining errors are our own!).
— Our Publisher, Springer, notably Executive Editor Ramon Khanna and Editorial Assistant Christina Fehling.
— Utrecht University, and especially Bert Theunissen, for supporting David's research both morally and financially.
— Larissa and Elinor for providing delightful distraction to David.

– Last but not least, our ever forbearing wives, Birgitta Nordström, Hermione
  Giffard, and Asta Petersen. ☺

Many people and institutions have provided pictures for selection in this
book. Their identity is given in the captions, and we thank every one of them
for their willingness to make these images available for the book and thus for
its readers.

Introduction

The International Astronomical Union (IAU) turns 100 years in 2019, and this book is written to celebrate that event. Many popular astronomy books recount the historical evolution of the fascinating science of astronomy, from the earliest times to today's understanding of the world in which we live. Some books focus on specific projects, and quite a few deal with individual scientists. The progress of astronomy has many parents, including not only the stubborn pursuit of good ideas, the advance of technology, and meticulous and painstaking work, but also the support of funders and enthusiasts. Astronomy has a unique power to fascinate laymen and professionals alike, to spellbind children and adults, and to foster curiosity and the urge to embark on a voyage of discovery—the greatest that is still left for us, inhabitants of a small planet in the vast sea that we call the Universe.

At least as important was the institutional framework within which scientific research is conducted and which makes it possible. This includes universities and observatories, but for astronomy also the one international organisation that aims to unite the worldwide scientific community. As the IAU sets out to celebrate its centenary and plan for the future, this book describes its evolution and the role it has played in the scientific community. In doing so, it attempts to answer three central questions: *What* is the IAU for? *Who* is the IAU for? *What* has been its role in the intellectual fabric of world astronomy?

The IAU now has 83 national members and more than 13,500 individual members—a significant share of all professional astronomical researchers. It hosts conferences, coordinates the names of astronomical objects, and recommends standards that facilitate the exchange of data. But the IAU does much more than this. It also includes professional offices for astronomical outreach, for young astronomers and for 'astronomy for development'. It thus provides an international platform for scientific discussions and also actively promotes astronomy in many different ways. But behind the facts and numbers, personal contacts are still at the heart of the IAU's function. The story of the IAU is essentially the story of the people who made it happen, and the many interviews which form the backbone of this book show importance of these connections: a remarkable number of personal friendships and many research collaborations have originated at IAU meetings.

The IAU was founded in 1919 '*to facilitate the relations between astronomers of different countries where international co-operation is necessary or useful*' and '*to promote the study of astronomy in all its departments*'. These aims have led the IAU throughout the century of its existence, but the way it has tried to fulfil them has changed. The first aim has been a resounding success: the IAU General Assemblies have become major conferences, with up to 3000 participants—a significant fraction of the global astronomical community. The General Assemblies cover the complete spectrum of astronomy. The proceedings of the IAU Symposia, now approaching 350 volumes, allow specific topics to be addressed in depth, are also a testimony to the development of front-line astronomical research.

In the first period of its existence, one could describe the IAU as a parliament of national representatives, who came together to discuss issues of mutual interest. That has changed radically. Today, the IAU aims to *include* the entire research community, undivided along national lines, with individual membership in the Union and increasing attention to diversity and the crucial role of young researchers.

The IAU has constantly striven—and occasionally struggled—to protect international scientific cooperation across the deep political divides that characterised the twentieth century, while maintaining an important function in the context of the rapidly evolving science itself and the changing fabric of institutions involved in astronomy.

But the role of the IAU in the astronomical community has also changed. We will demonstrate how the emphasis of the IAU's activities has shifted from facilitating collaboration by organizing meetings and defining common standards to promoting astronomy within society at large. The Pluto controversy of 2006 and the successful International Year of Astronomy in 2009 were important turning points. The Union is now mostly aimed at the outside world: the IAU is the *representative* of the global astronomical community. In other words, its audience is no longer just astronomers, but the entire world population, from policymakers to school-age children. An ambitious and comprehensive Strategic Plan 2020–2030, setting this out in more detail, was approved by the 2018 General Assembly in Vienna and will be its guide for the future. In other words: the way the IAU is serving astronomy has changed fundamentally, although it has historically been characterised by great continuity. In this book, we will trace this history in the context of the turbulent scientific and political history of the past century.

This is the second book on the history of the IAU. In 1994, former IAU President Adriaan Blaauw (1914–2010) published his history of the IAU up to 1970, based on the IAU archives. Accordingly, his book focused mainly on the efforts of the IAU to adapt to the two World Wars and the Cold War. Blaauw also provided a detailed account of the discussions about the organisation of the IAU. We have made thankful use of his work, but we all had the common vision to write a broader book, which includes the context in which the IAU has operated. Thus, we adopt a different perspective: we attempt to describe the development of the IAU on the scientific, technological, and political background of its time. For that reason, our book also covers the period that he has treated so expertly. It also means that not all officials or meetings will be mentioned in this book and that there will be little attention to financial issues. We therefore emphasise that inclusion or exclusion does not reflect importance.

This book originated at a symposium commemorating Adriaan Blaauw's centennial in 2014. This gave former IAU General Secretary Johannes Andersen an idea, which he immediately discussed with David and Claus. The volume naturally focuses on the second half of the century, when the majority of the key players are available for interview. The text is our joint responsibility, with the following rough division of labour: Claus collected the interviews with former and current IAU officials and some other key figures, who offer their personal perspectives on the historical developments, while David focused on the institutional history and the political context, and Johannes wrote sections on the scientific developments. Throughout, Johannes has been untiringly the promoter of this project, although his interview in this book states that 'looking back is not in the nature of Andersen'.

Contents

1

The Birth of the IAU (1919–1929)

International Cooperation Before the IAU

The stars and planets are the same everywhere; we can just see different sections of the sky depending on our location on Earth. For that reason, astronomers have always wanted to compare data and insights. Eclipse predictions, ephemerides and star catalogues were sometimes kept secret, but more often they circulated, for others to use and improve. This quantitative type of data is relatively easily shared across borders. Astronomers themselves travelled internationally too: Tycho Brahe's observatory on the island of Hven attracted many young astronomers to Denmark at the end of the sixteenth century. Some stayed for a few years and brought Tycho's innovative methods and instrument designs along with them as they returned home. A few years later, in 1608–1609, news of the new telescope, and Galileo's ground-breaking discoveries made with it, spread quickly.

© Springer Nature Switzerland AG 2019
J. Andersen et al., *The International Astronomical Union*,
https://doi.org/10.1007/978-3-319-96965-7_1

A pertinent trend in astronomy has been that its practitioners converge around research infrastructures and in international collaborations. Large research facilities attract scholars as well as decision makers, such as in this picture, where Tycho Brahe receives King Jacob VI of Scotland (later also King James I of England) at his Uraniborg observatory on Hven. Courtesy of the Royal Library, Copenhagen

These contacts crossed borders, but could not yet be described as 'international cooperation'. The contacts were personal, not institutional, and not between representatives of nations. The Venus transit observations of 1761 and 1769 initiated international cooperation in a more organised way. Several expeditions were sent out across the globe to observe these phenomena, with the explicit intention to compare their findings, in order to establish the distance of the Earth to the Sun. But the expeditions were still mostly organised along national lines, funded by governments or monarchs. In this period, most major astronomical institutes were royal or national observatories, of which many were founded in the eighteenth and nineteenth century, often related to maritime ambitions.[1] They also cooperated in projects to determine the longitude difference between them, and they routinely exchanged publications.

In the nineteenth century, astronomers, like scientists in other disciplines, started organising themselves in professional societies, meeting at regular conferences, and publishing journals that were circulated internationally. In 1820, the (British) Royal Astronomical Society was founded, which published the *Monthly Notices*. It was followed in 1863 by the German *Astronomische Gesellschaft*, which actually had members from many countries, including non-German speaking ones. It published the *Astronomische Nachrichten*. The *Astronomische Gesellschaft* also organised a central service to quickly spread news about astronomical phenomena: the Central Bureau for Astronomical Telegrammes (1884), which was managed from Kiel in Germany.[2]

Quiet before the storm: Attendees of the Astronomische Gesellschaft conference in Hamburg, in August 1913. Only one year later, international meetings in Germany would be impossible. Photo: University of Chicago Photographic Archive Library, apf6-04479r, Special Collections Research Center

---

[1] Dick (1991).
[2] Pfau (2000).

International cooperation in astronomy was facilitated by the introduction of a new technology that revolutionised how astronomers worked: photography. The photographic plate could initially only be used for very bright objects such as the Moon, but from the 1870s it became possible to photograph stars and planets. Especially amateur astronomers such as Sir William Huggins and his wife Margaret played a large role in introducing photography in astronomy. By the end of the century, it became the main detector in astronomy, until supplemented by the photomultiplier after the Second World War. Its quality and sensitivity was greatly improved over that period.

What made photography so revolutionary was, first, its ability to integrate faint light over long periods, instead of the few milliseconds of the human eye, and thus to detect much fainter sources than before. Other revolutionary features were the supposed permanent and impersonal character of a photograph, and finally its transportability. Once developed and fixed, a plate became a permanent record of the state of that part of the sky at the moment of the observation, which could be measured by anyone at any other time. This enabled astronomers to separate the task of observation from that of its (later) evaluation. In this way, Jacobus C. Kapteyn in cloudy Groningen could measure plates from the Cape Photographic Durchmusterung that the Scottish astronomer David Gill sent him from the observatory at the Cape of Good Hope.[3] Some American and European observatories set up telescopes in the Southern Hemisphere, especially in South Africa, which returned photos for measuring. In this way, the rich Southern skies could be explored efficiently, even though most astronomers worked in the Northern hemisphere.

Moreover, repeated photographs of the same field enabled others to study changes in the positions or light output of stars and other astronomical sources, and in the case of variable stars determine their period(s). This led to Henrietta Leavitt's epochal discovery of the period-luminosity relation for Cepheid variable stars in the Magellanic Clouds, which were all at the same distance. This, in turn, made distance measurements possible for stars beyond the reach of other methods (trigonometric or statistical parallaxes). After it became possible to capture the spectra of hundreds of thousands of stars on photographic plates, Annie Cannon and her colleagues at Harvard College Observatory classified all the stars in a new system, based empirically on line strengths and colours, and indirectly related to surface temperatures. The resulting Henry Draper (HD) catalogue was a memorial to Draper himself and his wife, who supported the project financially, but also to the organisa-

---

[3] Van der Kruit (2014).

tional and fund-raising genius of the observatory director, Edward C. Pickering. All talents were needed to complete this monumental project!

Impressed by the possibilities of photography, a group of prominent astronomers met in Paris in 1887 and decided to embark on a coordinated international project to divide the entire sky from pole to pole into zones and photograph it. The result would be the *Carte du Ciel,* an all-sky photographic atlas of the nearby Universe, freely available to all, and a matching *Astrographic Catalogue* of measured positions of all stars. This was a truly international undertaking, with 20 participating astronomical institutes from Europe as well as from Argentina, Mexico, South Africa, and the British colonies India and Australia.

A project like this required much coordination. A 'parliament' was created to agree on instruments, methods and standards.[4] It was decided to use (near)-identical refracting telescopes with an aperture of ~300 mm and a focal lengths near 11 metres (resulting in a plate scale of one arcminute per mm). The magnitude and complexity of the project had, however, been underestimated, and unforeseen political events in the form of wars, economic hardships, and consequent shortages of staff and money had intervened. The work was finally terminated in 1970, when the *Carte* was still unfinished. The catalogue was completed, however, and its data has recently been integrated in modern astrometric projects.

The first almost global science project in astronomy was the Carte du Ciel project. These are the participants at the starting meeting in Paris in April 1887. Photo: Courtesy of the Bibliothèque numérique/Observatoire de Paris

---

[4] Lamy (2011).

The *Carte du Ciel* project demonstrated that common standards were needed, but also that it was hard to agree on them. The nineteenth century nevertheless saw many conferences on standardisation, most famously on the metric system. The need for standardisation increased with the increase of international travel and communication, but also very practically to coordinate boundary-crossing infrastructures such as railway and telegraph networks.

Already in 1875, common international standards for length and weight measurements (the metre and the kilogram) had been entrusted to the *Bureau International des Poids et Mesures* (BIPM) in Paris. Another important case of standardisation was time. Keeping track of time has traditionally been an astronomical task, related to the rotation of the Earth and its orbit around the Sun. Greenwich Mean Time had become a common standard by the end of the nineteenth century, but by the turn of the century it was realised that the position of the Earth's axis was not fixed with respect to the solid body of the Earth itself, thus complicating the problem for all and involving both astronomers and geophysicists. An international treaty establishing a *Bureau International de l'Heure* (BIH) as a global centre for time-keeping was essentially ready for ratification when the First World War suddenly broke out, putting the plans on hold.

For astronomy itself, there was no centralised organisation. The *Astronomische Gesellschaft* was international in principle, but mostly focused on the European continent. In 1904, George Ellery Hale founded a new international organisation: the International Union for Cooperation in Solar Research, known as the Solar Union, which despite its name also covered stellar astronomy and astrophysics. In 1910, this organisation had adopted the Draper classification of stars as the new standard, for example—until then, many different classification systems existed next to each other.[5] Hale was known both as a pioneer of astrophysics and as a hyperactive organiser, who (co-) founded several observatories, including those of Yerkes and Mt. Wilson, as well as the Astrophysical Journal and the American Astronomical Society.

---

[5] DeVorkin (1981).

George Ellery Hale at the spectrograph of the 60-foot Solar Tower Telescope at the Mount Wilson Observatory. Photo: Courtesy of the Archives, California Institute of Technology

Conferences were social events as well as scientific ones, and this extended beyond the meetings themselves. Lively scientific discussions took place aboard the ship to a meeting of the British Association for the Advancement of Science in South Africa in 1905, for example. Similarly, in 1910, the international astronomical community gathered in Cambridge, Massachusetts, for the meeting of the American Astronomical Society, then travelled by special train to Pasadena, California, for a Solar Union meeting, altogether spending several weeks together.[6] Compared to today, conferences were long, intense gatherings of selected senior researchers.

The conference in Pasadena also provided an opportunity to admire the new 60-inch reflector at Mt. Wilson, one of the observatories founded by

---

[6] DeVorkin (1981).

Hale. It was the biggest telescope in the world, later surpassed by the 100-inch refractor, also at Mt. Wilson. These telescopes, equipped with photographic plates, surpassed any nineteenth-century instrument because of their size, but also because the 'seeing' (image sharpness) at Mt Wilson was much better than any traditional city observatory in the Western world. From now on, telescopes were increasingly built in places where the atmospheric conditions were best, rather than near academic institutions. Moreover, photographic plates made it possible to 'transport' observations from remote areas to scientific centres. Combined with the sensitivity of photographic plates, this translated directly into not only finer detail, but also in detection of much fainter stars.

The crowning achievement, the 100″ telescope itself, combined a huge light collecting area with a superb site and revolutionised astronomy several times. Using it, Edwin Hubble determined the distance to the Andromeda nebula, using the Leavitt period-luminosity relation, and proved that it was far outside the Milky Way. Hubble also used the telescope to show that the Universe is expanding, and Walter Baade used it to establish the concept of stellar populations (see below). The lesson was that the quality of site is as important as the size and design of its telescope(s). European astronomers had long complained about the 'unfair advantage' of the Californian astronomers with their large, privately funded telescopes, and excellent sites—before they started to develop new telescopes at even better sites in the Southern Hemisphere.

Scientific Unions and Interwar Politics

The First World War, with its immense human and material toll, was traumatic for many reasons. Apart from the many personal dramas, for scientists it also meant a near-complete breakdown of international cooperation. Conferences were cancelled as travelling became impossible—the Solar Union never convened after 1913. Even mailing journals and letters became troublesome. But more importantly, many scientists from both warring sides became involved in the war, both the 'real' war and the propaganda war. In October 1914, a manifesto signed by 93 German intellectuals, including many scientists, supporting their country's army, caused so much anger abroad that all contacts—including scientific—between the Allies and the Central powers were terminated. The wounds inflicted by the atrocities of the war were so deep that relations were not re-established until long after the war. This had a severe impact on the first decades of the IAU.

The founding of the IAU was directly related to the war. Again, George Ellery Hale played a key role, not in his capacity as a leading astrophysicist, but as the chairman of the American National Research Council during the war. He wanted to create an Inter-Allied Research Council, pooling the research capabilities of the allied countries to support the war effort. After the end of the war, this plan led to the new International Research Council (IRC), founded in Brussels in 1919, which functioned as the umbrella organisation for launching several Unions in specific disciplines—Hale himself was actually not present. The International Astronomical Union and the International Union for Geodesy and Geophysics (IUGG) were the first ones, representing disciplines with relatively strong traditions of international cooperation. They were founded at the same meeting as the IRC itself, and their statutes served as models for the others.[7] Benjamin Baillaud became the first President of the IAU. He was director of the Observatoire de Paris and a very active organiser of international cooperation, especially as applied in time keeping and longitude measurements.

The US delegation members aboard SS. Aquitania on the way to the founding meeting of the International Research Council (and the IAU) in Brussels. Top row, standing from left: Frank Schlesinger, Walter Adams, William Wallace Campbell, Samuel Mitchell, Frederick Seares, Charles St. John. Seated from left: Benjamin Boss, Elizabeth Campbell, Dorothy Mendenhall. The photographer was Joel Stebbins. Photo: Courtesy of the University of Wisconsin-Madison Archives (ID 2018 s00068)

---

[7] Greenaway (1996).

Joel Stebbins, professor of astronomy and observatory director at the University of Illinois, and a keen photographer himself, is here pictured during a visit by the US IRC Delegation to the devastated WWI battlefields in Flanders and France. Photo: Courtesy of the University of Wisconsin-Madison Archives

The devastation of parts of Europe that Stebbins and his colleagues witnessed formed a stark background for the decisions about membership of the IRC and, by implication, the IAU. This photo, obtained by Stebbins, shows the ruins of the Cloth House of Ypres in Flanders, the location of fierce battles between the warring parties. Photo: Courtesy of the University of Wisconsin-Madison Archives

The first Unions were soon joined by unions for physics (IUPAP), chemistry (IUPAC), biological sciences (IUBS), mathematics (IMU) and the new radio science (URSI). All Unions were subject to fairly strict rules set by the IRC. The most important one concerned membership: the former Central powers were excluded. This included to Germany, a scientific superpower. After some hesitation, neutral countries were invited to join. Some, including the Netherlands and Scandinavian countries, accepted only reluctantly. They immediately started lobbying for full international membership, including Germany, but the resistance was strong. French and Belgian scientists in particular could not forget the horrors and war crimes of the Great War.

Soon, German membership became a divisive issue within the IRC and the Unions. It even caused the termination of the activities of the International

Mathematical Union in the 1930s.[8] It did not get that far in the IAU, but Adriaan Blaauw has described how the issue dominated the first decade of the Union. Jacobus Kapteyn from the Netherlands, for one, refused any connection with the IRC or its Unions as long as they were not 'truly international', as he put it. This meant that his *Plan of Selected Areas*, an international observational project that was intended as a more efficient alternative to the *Carte du Ciel*, remained outside the IAU, much to Hale's disappointment. Kapteyn's student Willem de Sitter became one of the most active lobbyists within the IAU for ending the boycott, together with his Danish colleague Elis Strömgren. The latter acted as a mediator between the IAU and the German astronomical communities, especially since he was also President of the *Astronomische Gesellschaft*. Later, Arthur Eddington served alternately in the Executive Committees of the IAU and the AG.

In 1925, many (but by no means all) IAU members signed a petition supporting lifting the ban, but in the end, they had to wait for the IRC, which in turn waited for the politicians. The IRC made its membership dependent on the League of Nations, so when—after the Locarno conferences in 1925—the League invited the Central powers to join, the way was open in principle. Germany itself proudly rejected the eventual invitation from the IRC, however; and in the end, it did not join the IAU until 1952. In the meantime, IAU presidents could (and did) make generous use of their prerogative to invite 'scientific men' [sic] from non-member countries to IAU meetings.[9] Thus, the General Assembly in Leiden in 1928 was attended by 14 German, 1 Austrian and 2 Hungarian participants.

In 1931, the IRC was reorganised as the International Council of Scientific Unions (ICSU).[10] The Unions thereby gained much more autonomy, and from then on, the IAU could make its own rules. Its new statutes terminated all restrictions on national membership, but Germany could still not join, now for financial reasons.[11] The new rules also formalised individual membership of the IAU (see below), by giving individual members voting rights.

The voting power and financial contribution of the member countries were initially based on population. Since colonies were included, colonial powers such as Britain and France were dominant. In fact, up to the Second World War, the Executive Committee, the governing board of the IAU, always

---

[8] Greenaway (1996) and Letho (1998).

[9] Blaauw (1994) 71.

[10] In 2018, ICSU merged with the International Social Science Council to become the International Science Council (ISC).

[11] Robert Wielen (S349 forthcoming).

included members from Britain, France and the US: the leading scientific nations as well as the winners of the Great War. Later, one seat was commonly reserved for the neutral countries. By 1938, the IAU had gone from 19 to 26 member countries, ranging from massive China and the Soviet Union to the tiny Vatican City State, which operated an active scientific observatory. It had 554 individual members, including some from non-member countries, who were co-opted by the Commissions.

In the 1930s, new political issues emerged, with the rise of fascism and the great purges in the Soviet Union. The latter particularly affected the IAU, when two astronomers disappeared and several others were threatened. The Soviet Union had joined the IAU in 1935—which was exceptional: it only joined ICSU and other Unions after Stalin's death in 1953.[12] The purges immediately created a dilemma for the IAU: to protest or not? On the one hand, can a politically neutral organisation interfere in political matters? On the other hand, doing nothing might also be interpreted as political—staying neutral sometimes requires speaking out. Since other Unions had the same problems, these kinds of issue were usually referred to ICSU, whose mission was to protect the interests of science in general. In this specific case, the IAU Executive Committee chose to remain silent, for the pragmatic reason that it feared that any protest would make matters worse for the Russian colleagues.[13]

Where Co-Operation Is Necessary or Useful

But let us return to the founding of the IAU. According to its statutes, the aims of the newly founded IAU were:

- *to facilitate the relations between astronomers of different countries where international co-operation is necessary or useful*
- *to promote the study of astronomy in all its departments*

As we saw above, the most heated discussions in the Union concerned the 'different countries' mentioned in the first aim: which countries (and astronomers) should be included? This has attracted most attention in existing

---

[12] Roberto Lalli (S349 forthcoming).
[13] Blaauw (1994) 121–127.

historiography.[14] In the meantime, however, the IAU's core scientific activities had a much more propitious start, with successful General Assemblies, a large number of active Commissions, and several active IAU-sponsored services such as the *Bureau International de l'Heure* and the *International Central Bureau for Astronomical Telegrams*. Despite the political controversies, the IAU's authority as the leading international astronomical organisation was uncontested. In practice, if not officially, it was a direct successor to the Solar Union, which was quietly dissolved.[15] It also benefited from the established reputation of the *Carte du Ciel*, which became connected to the IAU via the dedicated Commission 23.

The IAU was—and is—an organisation of *professional* astronomers. This may sound obvious, but it was not. Until the nineteenth century, it was hard to distinguish between 'professional' and 'amateur' scientists—some unpaid researchers made fundamental contributions to science. This was especially the case in astronomy.[16] The British Royal Astronomical Society, the German Astronomische Gesellschaft and the American Astronomical Society all counted prominent 'amateurs' among their members. Only in the late twentieth century did a PhD in astronomy become the standard entrance ticket to the astronomical community. The IAU is also an organisation for professional *astronomers*, not a joint organisation for astronomers and physicists, as Hale had originally wanted his Solar Union to be.

Like all Unions, the IAU was an inter-national organisation, in the sense that its members represented countries. Each country established a national committee which represented an 'adhering body' such as an academy of science or an astronomical society (countries themselves cannot be members since the IAU is a non-governmental organisation). In practice, the General Secretary of the Union maintained an address list of all the people who were involved with the Union, and communicated directly with them. This developed into individual membership, which was formalised in 1931, when individual members were given the right to vote in the General Assembly on scientific issues, while administrative and financial issues still had to be decided by national votes. This apparently happened without much discussion, simply confirming existing practice. And yet, it was exceptional: from then on, the IAU was the only scientific Union with individual members.

---

[14] Blaauw (1994), Greenaway (1996), see also for example Schroeder-Gudehus (2012).
[15] Blaauw (1994).
[16] Cf. Lankford (1997), Rothenberg and Williams (1999), Baneke (2010).

The direct involvement of individual astronomers may also be related to the number of scientific Commissions of the IAU, of which there were many more than in other Unions. This made it possible for many researchers to be actively involved. And active they were. The first General Assembly in Rome in 1922 was attended by 83 people; the second one, in 1925 in Cambridge (UK), was more than twice as big; by the third, in Leiden in 1928, the number of participants was 249 (including guests); the IAU had 288 members at that time. The fourth one was the first outside Europe: in Cambridge, Massachusetts, in 1932, timed to coincide with a solar eclipse. The USA was quickly becoming a great astronomical power, not least because of the new large telescopes on the West Coast and the rapidly expanding astronomy programmes at the East Coast universities.

A group of French astronomers travelling with the SS Lafayette to the IAU General Assembly in Cambridge, Massachusetts, in August 1932. Photo: Bibliothèque numérique—Observatoire de Paris

The triennial General Assemblies quickly became the most important international conferences in astronomy, attended by the elite of the field. They became the most visible manifestations of the Union. Organising scientific conferences was the main way in which the IAU 'facilitated the relations between astronomers of different countries'. Their prominence is illustrated by the fact that an astronomer's seniority is sometimes measured by the first General Assembly that he or she attended. Apart from the scientific discussions, important standards and definitions were prepared by Commissions and formally endorsed through resolutions of the General Assembly. Early examples include the standardisation of the three-letter designation of constellations, and the recommendation to use 'Universal Time' as defined by the Bureau International de l'Heure, instead of Greenwich Mean Time.

The General Assemblies consisted of some plenary sessions, including the actual 'general assembly' about the administration of the Union itself. They also included elaborate social events, including gala dinners, frequently with dancing. The opening ceremonies were often attended by heads of state or other high-ranking officials. In 1922, King Victor Emmanuel III of Italy attended the opening of the first General Assembly. In 1935, the President of the French Republic, Albert Lebrun, hosted a reception. During that conference, IAU president Eddington was reported to comment that stellar interiors were 'unlikely to be much hotter than Paris in July'(!).[17]

The participants to the 1922 General Assembly in Rome seized the opportunity of an excursion to Florence to admire Galileo's first telescope on exhibit there. Photo: The Historical Archive of the Arcetri Astrophysical Observatory

The scientific discussions mostly took place in the meetings of the 'standing committees', which later became known as Commissions. The founders of the Union proposed an initial list of 32 standing Commissions, corresponding to

---

[17] Reminiscence by McCrea in GA newspaper 1988, 4.

the fields in which they thought international cooperation would be 'necessary or useful'. An important task for the first IAU General Assembly in Rome in 1922 was to review this initial list and decide which Commissions would actually be created. Several proposed Commissions were eliminated before they were ever created, and some others were terminated within a few years. For example, the members of Commission 1 on Relativity, whose first president was no less a person than Arthur Eddington himself, decided in 1925 that 'collective enterprises cannot, at the present moment, improve progress in the field of General Relativity'; progress in this field was to be made by individual researchers. In this case, international co-operation was neither necessary nor useful; the commission was terminated.[18]

Other Commissions became very active, however. They were intended as active working groups, not as communities of researchers with shared interests. Initially, they typically had 6–12 members. Interestingly, members could also be recruited from non-member countries, thereby adding to the individual membership of the Union.

Because the Commissions were initially small, the membership of the union was highly selective. In practice, it consisted mostly of the most senior astronomers from Western countries. The overwhelming majority was male; women in higher academic positions were rare. Among the small number of women were prominent names, however, and several were active in multiple Commissions, including Cecilia Payne-Gaposhkin, Margaret Harwood, Charlotte Moore Sitterly, Marguerite L. d'Azambuja and Edmée Chandon. In 1938, Payne-Gaposhkin was the only female official, as president of a subcommittee of Commission 25 Stellar Photometry.

---

[18] Proceedings of the 1925 General Assembly, 13.

Cecilia Payne-Gaposchkin, chair of the department of Astronomy at Harvard University and president of a subcommittee of Commission 25. Photo: University of Chicago Photographic Archive Library, apf6-01303r, Special Collections Research Center

From the beginning, the Commissions were routinely identified by their number. In the course of time, their names have changed or been updated—sometimes informally—to reflect the evolution of their science. Vacant numbers were sometimes re-used if new Commissions were needed; otherwise the numbering system was simply continued. The Commission numbers have survived for nearly a century, from the initial list of 1919 until the thorough overhaul in 2015. (See commission history in Appendix B).

A key organisational issue for the newly-founded IAU was the governance (and funding!) of the *Bureau International de l'Heure* (BIH), which had been maintained by the French government during the Great War because the international treaty to establish it had not been ratified yet. But Benjamin Baillaud, the first IAU president, felt strongly that it was important to coordinate accurate time internationally. This led to involvement of the IAU, via a joint cross-union committee with the International Union of Geodesy and Geophysics (IUGG). In these early years, issues such as the definition of 'Universal Time' were major topics of discussion during the General Assemblies.

In the beginning, the director of the BIH was the only paid official on the IAU budget, via Commission 31 'Time'.[19] Until 1919, Baillaud was director; he was succeeded by General Gustave-Auguste Ferrié, an exceptionally gifted soldier, engineer, and organiser, and a graduate of the famous École

---

[19] Blaauw (1994) 6–7, 265–266.

Polytechnique. After the war, he founded the *Laboratoire National de la Radioélectricité*, where mixed teams of civilian and military radio physicists and engineers would work together on common problems. On this background, one is not surprised that he was heavily involved in both the IAU and the *Union de Radio Science Internationale* (URSI) as well as in the *Bureau International de l'Heure* (BIH), which transmitted its first accurate time signal from just the Eiffel tower. The BIH also calculated the exact difference between standardised 'radio time' and astronomically established time.

In addition, the IAU also sponsored the *International Central Bureau for Astronomical Telegrams,* an information service on new discoveries and transient phenomena in astronomy. It can be regarded as a new incarnation of the bureau of the *Astronomische Gesellschaft* in Kiel. Started in Uccle (Belgium), it moved to neutral Copenhagen in 1922, conveniently close to the central office of the AG, from where it started to publish 'Circulars'. The IAU was also involved in the *International Latitude Service* (ILS), together with the Geodetic union that had been founded by the *International Geodetic Association* in 1899 to study the precession of the Earth's axis, particularly its effects on measures of latitude.

Interwar Astronomy and the Rise of Astrophysics (1930–1939)

There was much astronomy to discuss at the General Assemblies and Commission meetings in these first two decades. Despite the constant political and economic turbulence, the Interwar years witnessed fundamental changes in our understanding of the Universe. The combination of new theories—especially general relativity theory and quantum physics—with the power of new large telescopes (especially those at Mt. Wilson) proved extremely fruitful. In these two decades the structure and size of the Milky Way were established, the first realistic models of the interior of stars were formulated, their power source was unveiled, and the expansion of the Universe was discovered.

A central figure in many of these developments was the brilliant British astronomer Arthur Stanley Eddington, the later IAU President during most of WW II. Eddington is probably best known to many for his spectacular confirmation of Einstein's General Theory of Relativity in 1919 and for his work and book on stellar structure. Starting as a mathematical prodigy—finishing first in the infamously difficult 'mathematical tripos' exam in

Cambridge, the first to do so after only two years of study—he became Plumian Professor and director of Cambridge Observatory.

Sir Arthur Eddington, photographed by Howard Coster in 1936. Photo: National Portrait Gallery

Eddington, also a Quaker and a conscientious objector during the War, was convinced that the theory of General Relativity was correct and was keen to verify the German physicist Einstein's prediction that the gravitational deflection of light might be measurable during an eclipse. Together with the Astronomer Royal, Frank W. Dyson (also later IAU President), he mounted an expedition to observe the Solar eclipse on May 29, 1919, which would take place in the Hyades cluster, in an ideal field rich in bright nearby stars. To a modern astronomer, their account of the reduction of the actual position measurements reads like a litany of horrors, but at the discussion at the Royal Society, the combination of Eddington's scientific eminence and Dyson's authority carried the day. Their work has been accepted as the experimental 'proof' of the theory of General Relativity ever since. Subsequently, the theory has been tested in many other ways, most recently by the detection of gravitational waves.

Up to the early twentieth century, the common term 'nebulae' comprised both gas nebulae and dark dust clouds near young stars, planetary nebulae around evolved stars, star clusters of all ages and sizes, and galaxies of all types—simply anything that looked blurry through a telescope. As the resolution and range of telescopes improved, the distinction between these very different types of objects gradually became clearer, and point sources within some of them became recognisable as stars. Some object remained 'nebulous' however. Were they 'true nebulae' or 'island universes', star systems that were simply too far away to be resolved?

Spectroscopy, which was introduced in astronomy in the last decades of the nineteenth century, revealed that the spectra of gaseous nebulae showed bright emission lines, while the spectra of normal cluster stars did not. As the Andromeda and most other spiral 'nebulae' also showed no strong emission lines, they probably consisted of stars, but their nature and especially whether they were nearby in the Milky Way or very distant objects was unclear. With time, large telescopes had begun to resolve a few of the brighter nebulae into stars and showed that they became much more numerous at the faintest limits. The nature of the Milky Way itself was also at stake: was it a spiral 'island universe' itself, or did it comprise the whole universe, including small spiral nebulae inside it? Both theories had been considered and rejected several times in the preceding decades.[20] Various methods were developed to solve this riddle and establish the size of the Milky Way once and for all.

William Herschel had started to use star count models to investigate the structure of the stellar system. From the late nineteenth century, this method was further developed by Kapteyn and others, who initially assumed that stars had equal luminosities at equal distances, and that the Sun was at the centre of a single star cloud. Interstellar absorption was assumed to not exist, or at least to be uniformly distributed. Of course astronomers realised that these were assumptions, but they needed something to work with. Gradually, star counting techniques developed into 'statistical astronomy'.[21] Harlow Shapley, one of the most prominent American astronomers of the twentieth century, chose a radically different approach. He used Cepheid variables as a distance indicator for globular star clusters, to enable studying their distribution in space. In 1918, he found that the geometrical centre of the Milky Way system was far away in the direction of Sagittarius, supporting a large Galaxy model. His estimated distances were roughly a factor of three too large, but the main elements of the picture were right.

There was another clue to the Galaxy's nature: the motions of the stars. As early as the 1910s, Eddington suspected that the systematic motions of nearby stars in the Galactic plane could be evidence of a spiral structure of the Galactic disc. This was strengthened by the Swedish astronomer Bertil Lindblad's studies of regularities in star velocities, which he interpreted in terms of a set of concentric stellar rings with regularly varying mean rotational velocities. This

---

[20] Smith (2006).
[21] Paul (1993), Van der Kruit (2014).

was taken up in the young Dutch astronomer Jan Oort's study of high-velocity stars. He found the values to be as expected if the Galaxy rotated differentially, not as a solid body, but with systematically varying angular velocity—analogous to the motion of the planets around the much heavier Sun. His results were soon confirmed by other astronomers, using different samples of stars. By the late 1920s, everyone was convinced that the Milky Way was a differentially rotating disc galaxy, with the Sun located far from the centre.

Gradually, the various methods converged, also because models of interstellar absorption of light were improved. The Milky Way turned out to be larger than most astronomers had thought, but spiral nebulae were separate systems. Important corroborating evidence came from Hubble's discovery of Cepheids in the Andromeda Galaxy, which made it possible to estimate its distance. This placed it firmly outside the Milky Way. The Universe proved to be far larger than anyone had thought.

The study of spiral nebulae marked the start of extragalactic astronomy: studying objects outside our own Galaxy. Astronomers soon focused their attention on measuring the properties of these new objects, above all their distances and radial velocities (towards or away from the Earth), which were already an established tool in studies of the stars of the Milky Way. Spectroscopy was already used to determine the radial velocities of stars (by using the Doppler effect), so the necessary tools were at hand. It remained a relatively small field for a while, though, in part because of the limited number of large enough telescopes. IAU Commission 28 changed its name from 'Nebulae' to 'Extragalactic Nebulae' only in 1948.

From a meeting in Leiden in 1923 with five pioneers of General Relativity: Albert Einstein, Arthur Eddington, Paul Ehrenfest, Willem De Sitter and Hendrik Antoon Lorentz. Photo: Leiden Observatory Archives

The possibility of a velocity-distance relation among 'nebulae' had been in the air for years, and Hubble's observations suggested that such a relation did indeed exist and that it was even linear, at least to a first approximation. At about the same time, the Dutch astronomer Willem de Sitter (the third IAU President), who had also worked on the cosmological consequences of Einstein's theory of general relativity, had derived a similar relation from the observations available to him (including Hubble's). The missing piece in the puzzle was the distance to more distant galaxies, which Hubble was in a unique position to determine, given his access to the 100″ telescope, the world's largest and most powerful. Using his earlier discovery of pulsating

Cepheid variables in nearby galaxies, he worked systematically and stepwise to determine galaxy distances through the magnitudes of their brightest stars, and finally from the brightness of the galaxies themselves. In 1931, he could construct a linear relation between redshift and velocity that convinced the large majority of astronomers.

Walter Adams, Sir James Jeans and Edwin Hubble in front of the 100″ telescope dome. Photo: Courtesy of the Archives, California Institute of Technology

This observational evidence was combined with a new model, proposed in 1927 by the Belgian priest and astronomer Georges Lemaître based on Einstein's General Theory of Relativity: the galaxies were not just flying away from us, the whole universe was expanding! In the cosmological puzzle, the pieces started to fall into place—an example of the growing links between theory and observation. The idea of an expanding universe was accepted by most astronomers in the early 1930s. Hubble himself was cautious to never identify the measured redshift with true velocities or Doppler shifts, however (Kragh and Smith 2003). (In 2018, IAU General Secretary Piero Benvenuti proposed renaming the Hubble Law as the Hubble-Lemaître Law, because Lemaître's original article already contained observational evidence as well as theoretical arguments. A resolution to this effect was subsequently submitted to electronic voting and passed by a strong favourable majority.)

As more distances of stars became known around the turn of the century, Ejnar Hertzsprung and Henry Norris Russell plotted them on the famous 'Hertzsprung-Russell diagram', demonstrating among other things the existence of dwarf and giant stars. Russell interpreted this diagram in terms of stellar evolution: the lifecycle of stars, which appear so unchangeable to us. At the same time, a few stellar masses and radii had been determined from eclipsing binary stars. These again underscored the enormous differences between dwarf stars and (super)giants, which were ultimately traced to the differences in structure and stage of evolution.

What the stars were made of, and what energy source that powered them, was still shrouded in mystery, however, even though Eddington published a ground-breaking model of the structure of stars in his famous and enormously influential book *The Internal Constitution of the Stars* in 1926. Understanding of the structure, energy source and composition of stars was made possible by the revolutionary developments in theoretical physics, starting with Niels Bohr's explanation of the atomic spectrum in 1913 and ending with the counterintuitive, probabilistic laws of quantum mechanics in the 1920s, which governed the behaviour of fundamental particles. For astronomy, the new theories meant that observed stellar spectra, colours, and luminosities, that had previously been used to classify stars on a purely empirical basis, could now be interpreted theoretically. They turned out to contain an astounding amount of information on the physical properties of stars and gas clouds.

As atomic physics and quantum mechanics developed in the first decades of the century, one began to glimpse what the Sun was made of. The chemical elements and their place in the periodic system were known, of course, but not their proportions in the Solar atmosphere—not to speak about the interior. In her PhD thesis, Cecilia Payne used the physical interpretation of the spectrum to derive the abundances of most elements, which she found to be remarkably constant from star to star and very similar to terrestrial abundances—except that she could hardly believe the high abundances she derived for hydrogen and helium (which later proved to be correct).

The new physics was taken up by many young astronomers, including for example Subrahmanyan Chandrasekhar and Bengt Strömgren. Chandrasekhar, an Indian astrophysicist who worked at Yerkes Observatory, used new physics to investigate stellar structures, for which he received the Nobel Prize in 1983. Strömgren, a prodigy whose father Elis was also a professor of astronomy and who became Denmark's youngest PhD at 21, was trained in classical astronomy, but he was greatly inspired by the work of Niels Bohr at the institute next door. He became one of the pioneers of the new version of 'astrophysics', meaning the application of advanced physical theories to astronomical problems. This became so common that 'astrophysics' is now sometimes taken as a more general and more prestigious term for astronomy!

Niels Bohr has been quoted as saying to Chandrasekhar, as late as 1932: '*I cannot be really sympathetic to work in astrophysics, because [...] you cannot tell me where the energy [of the Sun] comes from, so how can I believe all these other things?*'. Shortly after, in the late 1930s, Hans Bethe and Carl Friedrich von Weizsäcker showed that the main energy source was in fact nuclear fusion.

In the two decades between the First and the Second World War, our understanding of the universe changed quite fundamentally. The size and structure of our galaxy, the existence of other galaxies, the structure and the energy source of stars, and the baffling fact that the universe as a whole is expanding, were all established in this period. Astronomy may be the oldest science, but many central aspects of our current view of the universe are surprisingly recent—younger even than the IAU, of which the foundations were laid in these turbulent times, which were dominated by the aftermath of the Great War and its wide-ranging effects on entire countries, political systems, governments, and the world economy. However, even worse times were brewing.

The last pre-WWII IAU General Assembly took place in Stockholm in 1938. Here Jan Oort and Walter Baade are seen in a discussion during an excursion. Photo: Oort Archive/J. Katgert

2

Crises and Opportunities (1940–1959)

The Second World War and Its Consequences

The Second World War (WW II) was even more disruptive to international astronomy than the First. All communication between the warring parties ground to a halt. International contacts and travel became very complicated and soon largely impossible. There was no question of organising international scientific meetings anymore; the planned General Assembly in Zürich in 1941 had to be cancelled. Exchanging information, even journals, became difficult. Many countries were occupied or became war zones.

In the initial stages of the Second World War, scientific work could proceed in most places, but this soon changed in the war-affected countries. Scientists and scientific institutions were not spared; for example, the venerable and important Pulkovo Observatory, including its library and instruments, was utterly destroyed in the long and bloody German siege of Leningrad. Elsewhere, institutes were occupied, instruments were damaged or reassigned, and researchers were mobilised, forced into hiding, or worse. Most of them were in Europe, where many astronomical institutes were located, but the Asian war also affected astronomers and institutions in China, South-East Asia, and Japan itself. Even in places that were not directly touched by the war, work became difficult, because of the isolation and difficulty of getting new instruments and supplies.

Already before the war, many (especially Jewish) scientists had fled Germany; during the War many others tried to get out of Europe. In various places, private help was also organised for them. On a different level, Bart Bok organised a newsletter in which astronomical news was summarised and sent from the US to occupied Europe via neutral countries such as Sweden or

© Springer Nature Switzerland AG 2019
J. Andersen et al., *The International Astronomical Union*,
https://doi.org/10.1007/978-3-319-96965-7_2

Switzerland, in an attempt to maintain at least *some* international communication. After the end of the war, back issues of astronomical journals were distributed in order to catch up.

Only in one sense was the War a good time for astronomy: blackout rules meant that even within many cities the sky was pitch dark, which for example facilitated the German Walter Baade's ground-breaking research into stellar populations and Cepheids at Mt Wilson in the USA (where there was also less competition for observing time than usual!).

One would expect the mood in 1945 to be pessimistic. After the First World War, it had taken almost a decade to normalise international relations again, only to see a new series of calamities strike the international community in the 1930s and 1940s. The reaction after the Second World War was very different from the first, however. International organisations were quickly re-established, and new ones were founded, including the new United Nations (UN) and its organisation for Education, Science, and Culture, UNESCO. All countries could join; there was no boycott of former Axis powers, which in any case were now occupied by the victorious Allies. As we will see, the IAU also resumed its work with renewed energy, soon including—for the first time!—scientific heavyweight Germany.

The world changed dramatically in the first five years after the war, however. The United States and the Soviet Union had emerged from the war as the two undisputed superpowers, but the relation between them quickly deteriorated into the Cold War: a tense stalemate, neither peace nor all-out war. Europe soon became divided between them in an American and a Soviet sphere of influence, with defeated, war-ravaged Germany split down the middle. East and West took radically different courses, with Eastern Europe reorganised along Soviet communist lines, and Western Europe rebuilt under American influence, and with major American funding. The Cold War became the dominant geopolitical conflict in the next decades, influencing all international relations, including in science.

Another important development in the immediate post-war period was the independence of many former colonies, especially in Asia, including the giant India, which joined the IAU in 1946 (although initially there was some confusion about the adhering body[1]). Decolonisation was a painful, often violent process in many cases. The result was, however, that the international community of nations gradually became more diverse. The two Cold War superpowers jostled for influence in the new countries, although some fought hard to remain neutral. The geopolitical scene was further complicated by the Maoist forces' victory in China, turning the largest country on Earth (by population) into a communist stronghold, and leaving the small island of

[1] Dipankar Mallik (S349 forthcoming).

Taiwan in the hands of the opposing nationalist Kuomintang party. The representation of China would become a sensitive political issue in all international organisations, including the IAU.

The Cold War had a profound impact on science, including astronomy. Sharing information or travelling to meetings across the Iron Curtain became difficult, and travel was always closely monitored. Obtaining visa could be very challenging. Still, there were many examples of successful international exchanges, and IAU conferences could always be attended by scientists from both sides, although specific individuals could be denied permission to attend by their governments, and it could be difficult to obtain a visa. In general, however, IAU meetings were neutral forums where researchers from all countries could meet. This, indeed, was its main function in this period. It was an obvious way to 'facilitate relations between astronomers of different countries'—and no mean feat in times of geopolitical conflict.

The consequences of the wars were not solely negative, however, especially for astronomy. Military technologies such as radar, rockets and electronic computers, which were first developed during the Second World War, would radically expand the possibilities for astronomers. Astronomy might seem unlikely to be affected by politics, having few practical or military applications, but since the Cold War was also a war of prestige and propaganda, all fields were included. Besides, many astronomical instruments were closely related to military technology. Advanced radio technology, infrared-, x-ray- and gamma detectors, electronic computers, and especially space technology and space missions to the Moon and other planets are examples.[2]

More generally, the great powers kept funding scientific research and instrument development generously during the Cold War, creating a lot of opportunities for new research. Research staff and student numbers at universities increased dramatically, increasing the research work force and mobilising new talent from social backgrounds who probably would not have been able to attend university before. Physics was the main beneficiary, but the number of PhD's in astronomy also rose significantly, especially in the USA after Sputnik.

The new technologies and the growth of science changed the way research was organised. Before the Second World War, astronomy was a fairly small-scale science. Some telescopes might be large, but research was still mostly done by individual astronomers, who often assembled their own observations. Theoretical work too was done by individuals, sometimes cooperating in ad–hoc constellations. But during the war, many astronomers were mobilised into industrial-scale, military research projects (for example on radar, sonar or bomb sights), working in teams that functioned within unfamiliar, elaborate

---

[2] Cf. DeVorkin (1992), Smith (2011).

management structures. After the war, state-funded 'big science' projects became common in non-military research as well. Gradually, this way of working also entered astronomy. The large 200-inch Hale telescope that was opened just after the war was still funded and operated by one institute, but many later large astronomical instruments were developed by consortia of multiple institutes, often international ones, and funded at least indirectly by governments. Using these instruments required the combined expertise of teams of astronomers, physicists and engineers. This was especially true for new branches of astronomy such as radio astronomy and space-based research.[3]

The inauguration of the 200 inch Hale Telescope on Mount Palomar in 1948. Photo: Courtesy of the Archives, California Institute of Technology

[3] Cf. DeVorkin (2000), McCray (2004), Sullivan (2009).

Radio: From Noise to Astronomy

Space research, using new rocket technology, promised great benefits for astronomy, but astronomers waited another decade or more before starting systematic observations. Similarly, the great progress in electronic equipment offered great opportunities to astronomy, but it was only gradually incorporated in astronomical practice. In the case of radio astronomy, developments unfolded much quicker, thanks to interested individuals and public funding. Within a decade, radio astronomy had become an established new field of astronomical research.

Already in 1932, Karl Jansky had reported that noise-like radio signals were coming from outside the Earth's atmosphere, probably the Galaxy. Later, Grote Reber built a receiver to investigate this radiation. After the Second World War, it turned out that radar technology developed during the war could be used to observe this radio 'noise' from outer space. This was the start of an exciting new research field. In the first few years, it was difficult to connect radio observations to the existing astronomical knowledge, but once the field took off, it went quickly. Radio observations opened a completely new window on the Universe, which until then had almost exclusively been observed in visible light.

Initially, the field was mostly developed by former radar engineers in Great Britain, Australia and the United States, who naturally focused on instrumental development. Indeed, in the first few years, the immediate usefulness of radio research and its relation to astronomy and the IAU was not clear, however. Most researchers were not astronomers, and their main interest was technical rather than astronomical. Besides, 'optical' astronomers could not immediately see what relation radio observations had to their research topics, and the angular resolution was so poor that it was difficult to identify radio sources with any previously known astronomical objects. This slowly improved in the 1950s.[4]

One specific astronomical use was clear from the beginning. When he heard about Reber's observations during the war, the Dutch astronomer Jan Oort immediately wondered if radio waves might be used for investigating the galaxy: radio waves would not be obstructed by interstellar dust. At his request, Hendrik van de Hulst calculated that there should be an observable spectral line in the radio region at 21 cm wavelength. And there was! After the war, they used leftover German radar equipment to start observing. In the USA and Australia, others did the same. The 21 cm line was first observed by the Harold Irving Ewen and Edward Purcell at Harvard in 1951, and soon

---

[4] Sullivan (2009), Munns (2012).

afterwards by groups in Australia and the Netherlands. In a demonstration of international cooperation, their findings were published together in the same issue of *Nature*.

The hybrid status of the field was illustrated by the fact that both the IAU and URSI, the International Union for Radio Science, founded dedicated Commissions in 1948: IAU commission 40 'Radio Astronomy', and URSI's commission for 'Extra-terrestrial Radio Noise', later also renamed 'Radio Astronomy'. The two Commissions long existed next to each other, with many members in common, and discussing similar topics, including such basics as terminology, definitions, and nomenclature. Gradually, a division of labour developed, with the URSI commission focusing on technology and the IAU becoming the platform for the scientific discussions.

Within a few years, radio astronomy became fully integrated in the astronomical discipline. Radio receivers were called 'radio telescopes', and radio physicists and engineers were recognised as 'radio astronomers'.[5] By the time of the General Assembly in Rome in 1952, several radio sources had been identified with known optical sources, including supernova remnants and some peculiar elliptical galaxies (the first sources were the Crab Nebula, Cassiopeia A and Cygnus A). The first maps of the distribution of interstellar hydrogen gas in the spiral arms of the Milky Way Galaxy were also presented; the 21 cm spectral line even allowed measuring radial velocities. It was a bewildering variety of observations. The impact of the new methods is demonstrated by the fact that Commission 33 changed its name from 'Stellar Statistics' to 'Structure and Dynamic of the Galactic System' in 1955: there was now a more direct way to investigate the Galaxy, apart from counting stars.

Apart from exchanging findings and coordinating their efforts, radio astronomers had another reason for seeking international cooperation: keeping certain wavelength ranges free from earthly interference. Since the birth of radio astronomy, the allocation of radio frequencies between commercial and scientific users—primarily broadcasters and radio astronomers—had been a contentious issue. A joint IAU-URSI-COSPAR committee was founded in 1960 to represent the interests of radio astronomy to the *International Telecommunication Union* (ITU; now a UN body): the *Inter-Union Commission on Frequency Allocation for Radio Astronomy and Space Science* (IUCAF). Protecting radio bands for research purposes was a prime example of protecting the interests of astronomy, an increasingly important task for the IAU, also including the fight against light and (later) space pollution.

---

[5] Sullivan (2009) 434.

The emergence of radio astronomy changed astronomy, and with it the astronomical community. It demonstrated how the IAU's constituency changed—and how the IAU itself had a role in defining the boundaries of the discipline. After a hesitant start, it became an integral part of astronomy, and Commission 40 soon became the largest IAU Commission. It was part of the reinvigorated IAU that emerged after the Second World War, in dramatically changed political and scientific circumstances.

Restarting and Expanding the Union

During the Second World War, there was little the IAU could do except wait for better times. Hosting conferences was clearly impossible, and international communication was difficult at best. The originally planned General Assemblies in Switzerland (1941) and the USA (1944, planned to celebrate the opening of the Palomar telescope) could not take place. IAU President Eddington died in 1944. IAU General Secretary Jan Oort was cut off from international communication in occupied Holland. Vice-President Walter S. Adams, director of Mt Wilson Observatory, took over as acting General Secretary and managed the financial affairs of the IAU for the duration of the war.

Walter Adams (left) and Harry Plaskett at Harvard Observatory. Photo: Courtesy of the Archives, California Institute of Technology

Just after the war, as soon as communication lines were open, father and son Bengt and Elis Strömgren invited the Executive Committee and representative astronomers from selected countries to meet in Copenhagen in early 1946, to discuss how to best re-start IAU activities. The rest of the Executive Committee had already elected Astronomer Royal Herbert Spencer-Jones to succeed Eddington as interim President, but three Vice-Presidents also wished to retire, so a new leadership was needed. A formal General Assembly had to be organised, and new arrangements were needed for the continuation of various key services to international astronomy. Fortunately, the original Swiss invitation to hold the next General Assembly in Zürich in 1941 was renewed in time for 1948. It made sense to start afresh in a country that had escaped the destruction of war.

Fifteen countries were represented at the 'rump' Executive Committee meeting in Copenhagen in March 1946.[6] Representatives of Germany and Japan were not invited, but after some discussion, it was decided that those countries would not be boycotted in the same way as after the First World War. Decisions on individual Japanese and German members, however, would wait for the investigations that were going on at various universities into their activities during the war.[7]

Bertil Lindblad, Director of the Stockholm Observatory in Sweden and IAU President 1948–1952. Photo: ESO Archive

---

[6] Blaauw (1994) 142–147.
[7] Blaauw (1994) 148–149.

Bertil Lindblad was nominated as new President, with Bengt Strömgren as General Secretary. Both were later formally elected by the General Assembly, and both Scandinavians were members of Commission 33. As a new tradition to permanently strengthen the corporate memory of the IAU, it was also decided that outgoing President Spencer-Jones and General Secretary Oort would continue to serve on the Executive Committee for an extra three-year term, as 'advisors'.

Some of the IAU-related specialised offices also needed attention. The Time and Telegram Bureaus in Paris and Copenhagen kept functioning during most of the war, but two activities in Berlin had to be relocated. Before the war, the Astronomisches Rechen-Institut in Berlin-Dahlem had been the most prominent institute for minor planet research. Its activities were now divided over several centres, including in Heidelberg and Leningrad, where ephemerides were calculated. The most important tasks, including the observation, identification, and naming of the increasingly numerous minor planets, were taken up by a new 'Minor Planet Center' (MPC) in Cincinnatti, Ohio (USA), directed by Paul Herget. It was later moved to the Smithsonian Astrophysical Observatory in Cambridge, Massachusetts. As the discovery and importance of minor planets increased—and the impact hazard began to be taken seriously after the turn of the century—the MPC became of central importance, as we shall see in later chapters.

Another German project had been the cataloguing of all available information on variable stars by Richard Prager in Berlin-Babelsberg. In 1936 Prager was dismissed by the Nazi regime; a few years later he was able to move to the United States and was given a position at Harvard College Observatory, where he died in 1945.[8] His work was continued, in part at the Sternberg Astrophysical Institute in Moscow, and in part at the Astrophysical Institute at Postdam. All these activities were supervised by IAU Commissions, although the IAU could provide little or no funding.

The external institutional context also changed. ICSU, the umbrella organisation of the scientific unions, resumed its activities after the war, but next to it, a new organisation found its place: the United Nations Educational, Scientific and Cultural Organization, UNESCO. It had a broader mandate than ICSU, and a much larger budget. For the IAU, UNESCO became an important sponsor of scientific and educational activities. This was welcome indeed, because the IAU had great ambitions to expand its activities, and external funding was needed for this.

---

[8] Blaauw (1994) 124–127.

Before the war, the IAU fulfilled its mission of 'facilitating the relations between astronomers of different countries' mainly by organising the triennial General Assembly, where astronomers from different countries would meet and discuss astronomical issues at Commission meetings. But already in Copenhagen in 1946, two new Commissions were founded with aims that were very different: to stimulate international cooperation in various different ways. These Commissions indicated the roles the IAU would wish to play in postwar astronomy. Henceforth, the Union saw a task for itself in actively supporting individual astronomers.

Frederick J.M. Stratton, the long-time pre-war General Secretary of the IAU, here seen during a Solar eclipse expedition to Japan in 1936. Photo: University of Cambridge/ Institute of Astronomy

The first new Commission formed at Copenhagen was no. 38: 'Exchange of Astronomers'; its president was former IAU General Secretary Frederick J.M. Stratton. As its name suggests, it supported international exchange visits for young astronomers to experience the inspirational influence of another, foreign, institute. Fortunately, the newly-founded UN organisation UNESCO could help with the necessary financial resources, and for several post-war years, the IAU filled a real need in this regard. This made it possible for the young French astronomer Jean-Claude Pecker to spend a year in Utrecht, for example. Pecker would be a most prolific IAU activist during the following half-century or more. He was soon followed by others. All in all, 70 exchanges were supported during the first ten years.[9]

The Executive Committee also decided 'That Dr Shapley be invited to emphasize to the United Nations Educational, Scientific and Cultural Organization (U.N.E.S.C.O.) that [the IAU] envisage the foundation of one or more International Observing Stations at which astronomers of all countries could be enabled to use instruments that are beyond the possibilities available in most countries. It was emphasized that such an observatory, internationally planned, internationally staffed and internationally operated, would effectively promote good will among scientists the world over, and thus serve the cause of peace.' The realisation of this plan was delegated to the new Commission 39 'International Observatories', an initiative of the Polish delegation (consisting of Tadeusz Banachiewicz and Jozef Witkowski) as well as of the Director of Harvard College Observatory, Harlow Shapley, who became the commission's first president.[10]

Sharing telescopes and instruments was a novel idea. Until then, telescopes were owned by individual institutes. Some collaborative arrangements did occur—for example, Leiden Observatory operated a telescope in Johannesburg in cooperation with the Union Observatory there. Travelling astronomers also routinely used each other's instruments. However, it became clear that, in the hard post-war years, nobody could compete with the giant Californian telescopes, especially with the new 200-inch Hale telescope at Palomar Observatory.

The obvious solution was to pool resources to build joint instruments. This was made easier by the fact that telescopes were increasingly located at remote places, chosen for their seeing conditions, rather than proximity to a research institute. This was a clear case in which international co-operation would be 'necessary or useful', and for the IAU to become involved in this endeavour thus appeared perfectly logical.

Pooling resources to build large telescopes indeed became common practice, but in the end this became managed by dedicated national or even intergovernmental organisations, not by the IAU. Two of the most prominent were the American Association of Universities for Research in Astronomy (AURA), and the European Southern Observatory (ESO). Both planned to operate telescopes in the Southern Hemisphere, although it took many years

---

[9] Reported in GA newspaper 1958, 6.
[10] Proceedings of the 1948 General Assembly 27–29; cf. Blaauw (1994) 147.

before the instruments in Northern Chile were completed. ESO was modelled after the successful European Centre for Nuclear Research, CERN. Jan Oort, Otto Heckmann and several other prominent IAU icons played key roles in founding it.[11] From the 1960s, the American and European space agencies NASA and ESRO (later ESA) also coordinated the international efforts to build space-based observatories (see next chapter). International cooperation became the norm for large ground- and space-based instruments in the late twentieth century.

As indicated, Commission 39 and the IAU did not manage to play any significant role in the development of international telescopes; the commission was terminated in 1955. One reason was probably that the IAU was unable to raise the necessary funding: governments or private foundations preferred to fund specific telescope projects directly. Indirectly, the IAU probably did play a role by facilitating international contacts and by providing a platform to discuss plans for future facilities. Formal discussions took place in Commission 9 'Astronomical Instruments' and at Symposia and Colloquia, but just as importantly, many ideas were discussed informally, in the margins of IAU meetings. The first ideas about ESO were discussed during IAU Symposium 1 in Groningen, for example, and more than one successful international cooperation started over drinks or dinners.

The IAU in the Early Cold War

The General Assembly in Zürich in 1948 was an occasion to mark the re-birth of the IAU and look back (Commission 41 'History of Astronomy' was created here). Future General Secretary Edith Müller, a Swiss astronomer, later compared it to a 'family reunion'. She also recalled that the excursion to the Sprüngli Chocolate factory was an especially popular event, since most attendees had not tasted chocolate for many years.[12]

---

[11] Blaauw (1991), Madsen (2012).
[12] Müller's Reminiscences in GA newspaper 1988, 4.

The change of guard. Bertil Lindblad, Harold Spencer-Jones, Jan Oort, and Bengt Strömgren photographed at the 1948 IAU Meeting in Zurich. Photo: Oort Archive/J. Katgert

But of course the meeting was primarily an occasion to take stock of the many changes that had happened in astronomy and in the world around it, and to make new plans. Radio astronomy promised to open a new window on the Universe. Nuclear physics had progressed quickly during the war and simultaneously led to new insights into the deepest interior of stars—and soon into the beginning of the Universe itself. The 200-inch Hale telescope in Palomar was nearing completion. Chile, Ireland, Finland, Hungary and just-independent India were admitted as national members of the IAU.

While the USA was clearly the leading country in astronomy after the Second World War, the IAU tried to remain neutral by maintaining a careful balance during the Cold War. For example, there was always at least one member of the Executive Committee from the US and one from the Eastern Bloc (in practice the Eastern member always came from the Soviet Union, Poland or Czechoslovakia), just as Britain, France and the US always had been represented on the Executive Committee in the Interwar years. That said, the Western countries were usually in a majority; relatively many officials were from France, the Netherlands and the United Kingdom, for example.

For the IAU, the biggest and most critical incident of the Cold War occurred already in 1951. It concerned the planned General Assembly in Leningrad of that year.[13] Before the war, the Soviet Union had invited the IAU General Assembly there, and the Soviet offer was accepted at the General Assembly in Zürich in 1948. According to Leo Goldberg, American astronomers had also considered an invitation, but they feared that astronomers from Communist countries would not be able to obtain a visa for the US in those years of 'the Red Scare'.[14] In the following years the political situation even deteriorated; this was the period of the great arms race and deep political crises. Most notably, in 1948–1949 the Soviet Union blockaded Berlin, and in 1950 the Korean War broke out. Correspondence in the IAU archives suggests that IAU officials were also worried about the anti-Western propaganda in the Soviet press, which attacked Western science and scientists, including 'bourgeois' and 'imperialist' astronomy. ICSU President Stratton (a former IAU General Secretary) also feared that astronomy in the Soviet Union might become subject to Party rule, as had infamously happened in agricultural science with the rise of Lysenkoism.

In March 1951, only five months before the General Assembly, the Executive Committee informed the Soviet Academy of Science that the meeting had to be 'postponed'. It stressed the non-political nature of the decision: '… the present international situation is likely to obstruct a representative international participation in the General Assembly and the Symposia'.[15]

---

[13] See Blaauw (1994), Chap. 8d.
[14] Goldberg, AIP interview (1983), session II.
[15] Quoted by from Blaauw (1994) 167.

Giorgio Abetti, here at the IAU GA in Zurich. Abetti—the son of Antonio Abetti, who as IAU Vice President (1919–1922) had played a vital role during the earliest phase of the IAU—secured the 1952 GA in Rome, as the original meeting foreseen to take place in Leningrad was cancelled. Photo: University of Chicago Photographic Archive Library, apf6-04366r, Special Collections Research Center

This case illustrates how hard it can be to maintain political neutrality. What would be the more political move, going or not going? The IAU was a politically neutral organisation, after all, but like it or not, it was forced to react to a political situation, and this was the position taken by the Executive Committee of the time, led by Lindblad and Strömgren. Fortunately, the worst did not happen in the end; the Soviet Union did not withdraw from the IAU, and Vice-President Viktor Ambartsumian remained in office. The Italian IAU Vice President Giorgio Abetti quickly stepped in and invited the IAU to Rome; and the General Assembly was held there in 1952, with only one year of delay.

At the IAU GA in Rome in 1952 (Left to right: Unidentified, Pieter Oosterhoff, Unidentified, André Danjon, Otto Struve, Harold Spencer-Jones, Georges Tiercy, Viktor Ambartsumian, Jan Oort, and Bengt Strömgren. Photo: Oort Archive/J. Katgert

The Rome meeting was notable for several reasons. Japan was re-admitted as a full member of the IAU, and Germany was admitted for the first time, represented by the *Astronomische Gesellschaft* (between 1964 and 1991, the membership was split between East and West Germany). The German-American astronomer Walter Baade presented his work on stellar populations and his recalibration of Cepheid-based distance measurements, doubling the size of the Universe. The first astronomically significant radio observations were presented. The conference is also remembered by many participants for the meeting with Pope Pio XII at the Vatican Observatory at Castel Gandolfo. At the advice of the astronomer and priest Georges Lemaître, the Pope carefully refrained from speaking favourably about big bang cosmology, as he had done a few months earlier.

Pope Pius XII greets Harlow Shapley following the Pope's address to the IAU visitors at Castel Gandolfo in 1952. Photo: AIP Emilio Segrè Visual Archives, Shapley Collection

By then, Pieter Oosterhoff had taken over as General Secretary, Bengt Strömgren having resigned in 1951 to become Director of the Yerkes Observatory. Oosterhoff was assisted by Miss Nel Splinter as permanent Administrative Assistant, a new position. This had become necessary because the workload for the IAU administration was increasing, for example with the launch of the successful—and still ongoing—symposium series, with Symposium No. 1 in Groningen in 1953 on 'Co-ordination of Galactic Research' (more on symposia in the next chapter).

The next General Assembly, in 1955, was in Dublin. The choice of this location in a small, neutral country was another compromise. By then, the international situation had changed again. After Stalin's death in 1953, a General Assembly in the Soviet Union become possible again. IAU President Otto Struve and Vice President Ambartsumian strongly argued for meeting in the two largest countries. Accordingly, the next two General Assemblies would take place in Moscow in 1958 and in Berkeley, CA in 1961. Outgoing President Struve from Berkeley, of the famous Struve astronomical dynasty and speaking in his mother tongue—Russian—paid eloquent tribute to the patience and forbearance of President Lindblad and General Secretary Strömgren, without which 'the Union might well have disappeared'.

Otto Struve, the Russian-born American astronomer, director of Yerkes Observatory, and IAU President. Photo: University of Chicago Photographic Archive Library, apf6-00157, Special Collections Research Center

The Moscow General Assembly in 1958 was a notable success, despite the recent uprising in Hungary in 1956, with more than 800 members attending. This was about three quarters of the IAU membership, which had risen above 1000 by this time. One of the most-discussed astronomy-related topics was the 'science of Sputnickery', as IAU veteran Shapley called it.[16] The first Sputnik had been launched in October 1957, only months before the General Assembly, during the International Geophysical Year. The launch took the whole world by surprise, not least the astronomical world—not because the idea of a satellite was unexpected (a launch had been expected soon anyway), but because it was the Soviet Union that launched one first, and a relatively large one too. It raised great expectations for future research possibilities. Appropriately, Commission 44 'Observations from Space' was created in Moscow. Shortly after the General Assembly, a Soviet space mission was able to secure the first-ever images of the previously unobservable far side of the Moon, underscoring again the Soviet prowess in spaceflight.

[16] GA newspaper 1956, 18.

From the IAU GA in Moscow, 1958: Left to right: Bertil Lindblad, Nicholas Mayall, Jan Oort, Mrs. Mayall (Kathleen Boxall) and Carl Seyfert. Photo: Oort Archive/J. Katgert

The General Assembly also featured the first daily newspaper: *Kosmos*, which contained articles in Russian, English, German and French. This example was to be followed at all subsequent General Assemblies. At the end of the General Assembly, Jan Hendrik Oort was elected IAU President. At the next General Assembly, in Berkeley, Viktor Ambartsumian (USSR) was elected President, with Donald Sadler (Britain) as General Secretary. Sadler had already been appointed Assistant General Secretary in 1957, a new position in the Executive Committee, created to prepare the General Secretary for their duties and enhance continuity in the Union's administration.

The IAU Executive Committee meeting in Herstmonceux in 1959, where the decision was taken to accept Taiwan as a member. Seen here (left to right) are Jan Oort, Olin Eggen, Richard Stoy, Richard Woolley, Petr Kulakovsky (interpreter) and Boris Kukarkin. Photo: Leo Goldberg, courtesy of AIP Emilio Segrè Visual Archives

Perhaps surprisingly to some, organising the 1961 General Assembly in Berkeley, USA, caused more political problems than the one in Moscow.[17] The organising committee, led by Leo Goldberg, had to make sure that all IAU members would get a visa. They even organised smaller advance meetings to test this. The main problem was that Taiwan had applied for membership of the Union. The (communist) People's Republic of China was already a member. Taiwan was not, but at this time, the USA officially considered Taiwan to be the only 'real' China. Excluding it would create big political (especially visa) problems for the General Assembly, but accepting Taiwan as a member would cause the People's Republic to withdraw. There was no neutral solution. Pressured by the American government, the Executive Board decided to accept Taiwan's application, though the decision was put on hold for a year in order not to have to accept it in Moscow, which would have created further diplomatic problems. The formal argument was that the Taiwanese bid conformed to all regulations, so it was impossible to reject, even though little astronomical activity was going on there at the time. As predicted, the People's Republic immediately withdrew; it would stay away for two decades, until a solution was laboriously negotiated around 1980 (see Chap. 4).

---

[17] This passage: Goldberg, AIP interview (1983), session II.

The Opening Ceremony of the IAU GA in Berkeley, 1961: Left to right: Ira Bowen, IAU President Jan Oort, US ambassador to the UN Adlai Stevenson, Leo Goldberg, Donald H. McLaughlin. Photo: Courtesy of the Oort Archive/J. Katgert

Visa for scientists from opposing blocs remained the most common practical problem caused by the Cold War. More sensitive were cases in which individual scientists encountered political troubles. Should the IAU support them? Again, there were cases in which remaining silent would be as politically charged as taking a stand. The IAU archives contain few documents on these issues, although they certainly happened. In most cases, the IAU referred politically sensitive issues to ICSU, whose mission was to represent *all* international science. Protecting free travel for scientists became a core concern for ICSU.[18]

Another way the Cold War influenced IAU meetings was by covert—and sometimes less covert—presence of intelligence officers at scientific meetings, and debriefings of scientists who had visited opposing countries. Both

---

[18] Greenaway (1996).

superpowers were greatly interested in each other's scientific and technological capabilities, including in fields that were close to astronomy, for example optical technology and radar. Eastern astronomers were also monitored by their own governments.

The Cold War also included occasions in which the IAU had to defend the interests of astronomy. An early example was Project West Ford, an American proposal to send billions of small copper needles (bipoles) into space to support military communication in the early 1960s. The IAU and other scientific organisations protested sharply against this 'pollution' of the pristine space environment—a topic that would return many times in the following decades, as space activities ramped up, both in the form of ever-increasing amounts of permanent space debris, and in the form of 'noise' generated at all wavelengths of the electromagnetic spectrum. Indeed, defending the interests of astronomy for all future in this practical sense became an increasingly prominent mission for the IAU.[19]

---

[19] Cf IAU Symposium 196 *Protecting the Astronomical Sky* (1999).

3

A Decade of Astronomical Surprises (1960–1969)

The Exploding Universe

In 1969, the 50th anniversary of the IAU was marked at two Symposia, in Basel and Brussels.[1] It was a modest way to celebrate, but astronomer's minds may have been elsewhere. The 1960s were among the most exciting decades in astronomical history.

In 1967, a military satellite registered an unexpected burst of gamma rays from space, which piqued the interest of the astronomical community. What kind of astronomical object could this be? In previous years, astrophysicists had already been surprised by many new discoveries. The diversity of celestial objects seemed bigger and stranger than anyone had thought only a decade before. But the 1967 discovery was not done with an astronomical instrument. In fact, the satellite had been designed to look down, not up, to detect gamma rays from nuclear weapons tests, which had been banned by the 1963 Test Ban Treaty. The astronomical discovery was a bonus, even though it was only published several years later.

This discovery is characteristic of the 1960s. Satellite technology and gamma ray detectors were initially developed for Cold War purposes, with funding on a scale beyond anything astronomers could hope for. But as it turned out, many of these technologies could be used for astronomy as well, and astronomers were needed to help developing them and interpreting the findings. In infrared observations, for example, celestial sources had to be

---

[1] IB 23.

© Springer Nature Switzerland AG 2019
J. Andersen et al., *The International Astronomical Union*,
https://doi.org/10.1007/978-3-319-96965-7_3

distinguished from satellites or incoming missiles, meaning that optical all-sky astronomical catalogues were crucial.

It is no exaggeration to say that astronomy exploded in the 1950s and 1960s. Every time a new wavelength range was observed, new and unsuspected kinds of objects turned up. This also meant that development of new instruments became an even bigger factor in the hunt for new astronomical knowledge than it already was. New technologies made it possible to observe new parts of the electromagnetic spectrum. First radio astronomy made the previously invisible long-wavelength Universe observable. Then the Space Race made it possible to observe X-ray, gamma ray and ultraviolet radiation that were blocked by the Earth's atmosphere. This yielded a staggering amount of new information and a dizzying series of new discoveries, including gas cloud structures, quasars, pulsars, gamma ray bursts, and background radiation. At the same time, probes to other planets revolutionised solar system studies. They culminated in 1969, when the first human expedition to an astronomical object reached the Moon.

If the development of astronomy in the first half of the twentieth century can be characterised as squeezing increasing amounts of information out of light by interpreting the optical spectrum of cosmic sources (read: stars) using new physics, astronomy after 1945 was characterised by the huge increase in the amount of radiation to be studied, both because a much wider spectrum became available for study and because detection methods became much more sensitive. After a period of great theoretical advances, astronomy became an observational science again. The Universe that this revealed turned out to be much more diverse, and much more dynamic, than anyone had thought.

The science of the Universe itself—cosmology—was no exception. Around 1930, the Universe had been discovered to be expanding, but at that time, no one had dared to speculate seriously about its beginning in physical terms. In the late 1940s, however, a group of physicists including George Gamow had started to think about a hot, dense early universe in which atomic nuclei were first formed. The theory of nucleosynthesis was further refined by Fred Hoyle and others. This was the start of 'big bang cosmology'. The term was coined by Hoyle, who ironically never believed in it himself, thinking that it violated fundamental philosophical principles. Instead, he advocated a 'steady state' universe that could expand without ever changing. Cosmology became of astronomical interest when extremely far (and therefore old) quasars were eventually identified, giving access to an earlier state of the Universe. For most scientists, the Cosmic Microwave Background radiation that was discovered in 1964 has confirmed the 'Big Bang' theory. Within these three decades, the known history of the Universe had radically changed.

The number of astronomers also continued to increase dramatically, thanks in no small part to increased government funding for science and technology. The astronomical community also became more diverse in the process, as astronomers were joined by physicists, computer specialists, and a wide array of engineers. Further, the new technologies changed the astronomers' way of working. Instrument development had always been an important part of astronomy, but now it became the main driving force of astronomical development.[2]

The astronomical scene had to adapt. The IAU was no longer the self-evident centre of the community. As a non-governmental organisation with a very modest budget, it had no natural role in instrument development. The schedule of the General Assemblies and their format did not seem to keep up with the developments. They did not expand in time—they already filled two weeks—but the schedule was filled with, at least according to some, unproductive Commission meetings. This caused a lot of discussions within the IAU about the way of working, the membership, and the format of the General Assemblies. Invited Discourses by eminent astronomers—in principle open to everybody—were introduced in 1955 as one way to make the General Assemblies more widely scientifically attractive.

---

[2] Harwit (1981), Smith (1997).

In 1963, IAU and URSI organised a joint Symposium on radio astronomy in Sydney. The meeting also included a visit to the new 64-metre Parkes telescope, where Edward 'Taffy' Bowen gave a tour of the dish. Jan Oort is second from left, then Ira Bowen, Viktor Ambartsumian, Gerald Mulders and Boris Kukarkin. Photo: Courtesy of the CSIRO Radio Astronomy Image Archive

In the meantime, scientific discussions flourished at the new IAU Symposia and Colloquia. These smaller-scale, more specialised meetings quickly became the main venues for discussing the latest developments in astrophysics. The series of published IAU Symposium proceedings are an impressive record of a rapidly advancing science. And with time, the International Schools for Young Astronomers (ISYA), started in 1967, came to signal a new, active role for the IAU in educating young researchers, especially in developing countries (see Appendix D). The increasing number of astronomical meetings was made possible by the fact that international travel became faster (and more comfortable) when commercial flights became more frequent and more affordable. Sometimes special charter flights from Europe or America were organised for General Assemblies.

Much of the technology (and funding) was related to the Cold War: this was the period of the great arms race and deep political crises, most notably the erection of the Berlin Wall in 1961 and the Cuban missile crisis in 1962. Pessimism and optimism went hand in hand—fear of nuclear war and expectations of unlimited atomic energy; suppression of democratic revolts (Hungary 1956, Prague 1968) and dreams of the Space Age; the Chinese Cultural Revolution and the Moon landings. In the previous chapter we have seen how the early Cold War affected the IAU. But as we see in this chapter, the Cold War also had some positive effects. Astronomy was one of the main beneficiaries of Cold War-related funding and technological development.

New Windows, New Views

The early years of radio astronomy and its integration into mainstream astronomy have been described in the previous chapter. In those first years, much research focused on the Sun, on meteors, and on hydrogen gas clouds in the Galaxy. The list of known radio sources also grew. The centre of the Galaxy turned out to be one, as well as the Crab Nebula, a well-known supernova remnant. Several other sources could not be identified with known objects, however. One of them had a particularly puzzling spectrum. In 1963, Maarten Schmidt managed to interpret it as a 'normal' hydrogen spectrum with an extreme redshift, meaning that the source must be further away (and more powerful) than anything observed before. Soon, similar other 'quasi-stellar objects' or quasars were identified. Understanding their nature became one of great questions of astrophysics. Eventually they were interpreted as accretion discs of supermassive black holes in the centres of galaxies.

The discovery of quasars was completely unexpected. One of the earlier breakthroughs in radio astronomy, the detection of the 21 cm hydrogen line (see previous chapter), was one of relatively few discoveries in astronomy that were predicted theoretically. One of the other great discoveries in the radio spectrum combined both characteristics. In 1964, Arno Penzias and Robert Wilson from Bell Laboratories struggled with 'radio noise' from all directions in the Universe. It soon turned out that several theoretical cosmologists, including Robert Dicke and George Gamow, had predicted precisely that. The Cosmic Microwave Background radiation was interpreted as a remnant of the earliest stages of the Universe, billions of years ago. The discovery of the background radiation was rewarded with the 1978 Nobel Prize in physics.

The discovery of the background radiation was a major coup in the ongoing controversy between two cosmological models, commonly described as 'big bang' versus 'steady state'. After this discovery, big bang cosmology became widely accepted, with only a small (and diminishing) number of dissenters left. Those dissenters were very vocal though: they kept arguing their case in many talks and publications, including at IAU General Assemblies. In 1970, IAU Commission 47 Cosmology was established. Fifty years and many space and ground-based experiments later, the Cosmic Microwave Background Radiation is still very actively and extensively analysed as one of the most informative diagnostics of the conditions in the early Universe.

A similar confluence of theory and observation occurred in 1967, when a British PhD student was working on a radio experiment to follow the faint interplanetary scintillation with high time resolution. Jocelyn Bell Burnell noticed a periodic radio source with no known optical counterpart that kept pulsing with a period of 1.33 seconds. It was initially jokingly called LGM-1, for 'little green men', but soon a second pulsating source was discovered in a different part of the sky. Many other 'pulsars' followed confirming the phenomenon but making intelligent life unlikely as a source. It took some time before they were interpreted as rapidly rotating neutron stars, predicted by Walter Baade and Fritz Zwicky as well as by the later IAU President Franco Pacini and Thomas Gold.

The discovery of pulsars was also rewarded with a Nobel Prize, though it was controversially awarded to Antony Hewish, Bell's PhD supervisor, but not to her. Ironically, Bell (who never complained) became more famous for not getting the prize than Hewish for getting it. In 2018, she was awarded a Breakthrough Prize 'for fundamental contributions to the discovery of pulsars, and a lifetime of inspiring leadership in the scientific community'.

Radio was the first new 'window' to be opened for astronomers, who had depended on optical light observations for many centuries. Many other new windows followed, however, thanks to the development of spaceflight. During the Second World War, German engineers, led by Wernher von Braun, developed the first modern rockets. In the last years of the war, V2 rockets were deployed against London and Antwerp. While they did not influence the final outcome of the war, they did attract the interest of scientists, engineers and military commanders alike. At the end of the war, the Allied forces captured many V2 rockets from Germany, bringing them home along with the engineers that built them. Apart from obvious military uses, they promised great astronomical potential: observations from outer space, unhindered by the

Earth's atmosphere. This meant that electromagnetic radiation that is normally absorbed or reflected by the atmosphere (and therefore never reaches the ground) could be detected. This included most of the infrared spectrum, as well as ultraviolet light, X-rays and gamma rays.

Initially, only brief observations of the brightest sources (especially the Sun) were possible. The 'sounding rockets' spent only a few minutes outside the atmosphere before falling back, and directing the cameras at specific points in the sky was difficult. Still, useful infrared, ultraviolet and X-ray observations were done. At the General Assembly in Berkeley in 1961, James Van Allen gave an Invited Discourse on one of the earliest discoveries made with satellites: the radiation belts that were named after him. The General Assembly journal also provided information on when the Echo communication satellite would be visible, an indication of the excitement of astronomers about space.

The first X-ray source other than the Sun was discovered in 1962 using an Aerobee sounding rocket. As it was the first of an unknown new type of object, and it was located in Scorpius, it was simply called Sco X-1. A few years later, it was identified by a dedicated satellite observatory as a neutron star that was a member of a binary system. This was the only type of object—apart from black holes—small and massive enough for any gas accreted from a companion to be heated sufficiently to emit the copious X-rays observed from this source. Riccardo Giacconi received a Nobel Prize for his pioneering work in X-ray astronomy.

More substantial astronomical observations required more time and a more stable platform than the sounding rockets could offer. This became reality during the International Geophysical Year of 1957–1958, when the Soviet Union launched the Sputnik satellite in October 1957. This was a major technological breakthrough as well as a propaganda coup of the first order. From now on, orbiting satellites carrying astronomical instruments were needed. The technological difficulties remained tremendous, however, as did the cost.

Initially, only the Soviet Union and the United States had the capability of launching satellites into orbit (the launchers were closely related to military missiles); France, Britain, Japan and China followed, but only in small numbers. Most scientific satellites were launched by the two superpowers; a viable European launcher, the Ariane rocket family, only became operational in the 1980s. Paradoxically, one effect of the superpower monopoly was that international cooperation in space science became common. In particular, NASA, the American space agency, and ESRO (later ESA), a collaboration of several European countries, became important drivers of international collaboration. In 1959, NASA offered to launch foreign scientific satellites for free, and offered technical assistance. It became common to combine instruments from several countries in one satellite. Because of their scale, cost and political sensitivity, space instruments were nearly always developed in close cooperation with government institutions.[3]

The first discoveries were followed by larger and longer missions that systematically scanned the sky in all wavelength ranges. As the technology developed and more satellite observatories were launched, a wide variety of sources of high-energy radiation were identified, from stellar chromospheres via white dwarfs, supernova remnants and neutron stars to galaxies with enormous, supermassive black holes. The gamma ray bursts mentioned at the beginning of this chapter demonstrate that non-astronomical satellites also contributed to the bonanza. 'High energy astrophysics' and infrared astronomy became significant sub disciplines of their own, as the phenomena were gradually better understood theoretically. In 1970, IAU Commission 48, 'High Energy Astrophysics', was founded.

The most spectacular space missions were not orbiting satellites, but probes that actually travelled to astronomical objects in the Solar System. The Moon was first, obviously. As early as 1959, a Soviet probe provided the first-ever photos of the far side of the moon. More missions followed, culminating in the six manned Apollo Moon landings in 1969–1972. Unmanned Solar System probes also visited other bodies in the solar system, including landings on Mars and Venus. Especially the Voyager missions to the outer planets captured the public's imagination as well as that of the scientists.

It seems to run deep in the human nature to assign names to things. The space missions provided an endless number of planetary and satellite surface features—an obvious task for the IAU. Since space was a major arena of Cold War competition, naming them became a sensitive issue. As planetary system exploration missions proliferated, lots of new planetary features were discovered that required naming. International coordination by a neutral organisation was required, and the IAU set to work; a central IAU Working Group for Planetary System Nomenclature and a Working Group for Small Body Nomenclature were formed to coordinate this.[4] Coordinating and standardising solar system nomenclature became one of the most visible activities of the IAU to the non-astronomical world. It still is (now including extrasolar planetary systems).

The immediately eye-catching discoveries came from new technologies, but the optical universe of planets, stars and galaxies remained the framework in

---

[3] Logsdon (1996), Krige and Russo (2000), Baneke (2010).
[4] Jana Tichá and Brian Marsden, 'A Short History of the Committee on Small Bodies Nomenclature', in IB 104 (2009) 72–73.

which the new objects were located and interpreted. Radio astronomy became integrated into astronomy after radio sources had reliably been identified with known optical objects. Optical astronomy drew less attention, however, because there were few spectacular discoveries to be reported—most had already been made long ago. Besides, the Palomar 200-inch (5 metre) telescope remained the largest optical telescope for several decades (a larger Soviet telescope was opened in 1967, but did not nearly have as much scientific impact, in part because of its location at a poor observing site). It was generally accepted that building larger primary mirrors required a different technological approach. In the meantime, several telescopes in the 4-metre class were being developed (see next chapter).

Early radio astronomy long struggled with poor angular resolution (sharpness), which made it hard to identify sources of small size at these long wavelengths. The solution was combining radio telescopes by means of interferometry, which is easier at longer wavelengths than shorter. It still requires extremely precise alignment and control of the telescopes, though, as well as complex data processing, so it was only possible with advanced electronics. Astronomer Martin Ryle shared the 1974 Nobel Prize for his contribution to this technology, which after overcoming the tremendous technical difficulties has later also been developed for infrared and later optical wavelengths.

The new kinds of telescopes did not only change astronomer's views on the nature of the cosmos; they also changed the way of working. A major new factor was the increasing role of electronics and computers—another World War 2 and Cold War heritage technology. Electronic and computer-based systems for telescope control were among the first astronomical applications, but increasingly also for detecting and processing the signals, and translating them into formats that astronomers could understand and interpret. Electronic detectors proved especially useful for automated photometric work (measuring the brightness of light sources).[5]

Radio astronomy led the way—radio interferometric observations and other data were all electronic rather than photographic, and all required extensive data processing. Radio astronomers were among the first major non-military users of computers. The staff of the Minor Planet Center in Cincinnati also used computers to calculate planetary perturbations on minor planet orbits, although fitting these calculations to actual observations was still laboriously done by hand.[6] Moreover, early electronic computers found extensive

---

[5] McCray (2014).
[6] Brian Marsden, 'History of the *Minor Planet Center*', IB 104 (2009) 67–71.

use in calculating stellar evolution models for a variety of chemical compositions, and for integration of stellar orbits in dynamical galaxy models of increasing sophistication.

The romantic idea of a single astronomer in a telescope dome, contemplating the Universe, became increasingly outdated. Electronically guided telescopes did not require hands-on operation, especially if the signals were also recorded electronically by computers. Astronomers could work in a heated, lit room rather than in a cold, dark dome—often in teams with engineers, data specialists, and other experts. Of course this development stirred nostalgia as well as excitement, and also some worries about the attitude of contemporary astronomers: an article in the General Assembly newspaper of 1970, reporting on a colloquium on Automation in Optical Astrophysics in Edinburgh, was headlined *'Automation No Substitute for Thinking'*.[7]

The military aspects of much of the technology could complicate things, but it also could provide opportunities. The large radio dish of Jodrell Bank near Manchester was not only an astronomical instrument, but also a showcase of technological prowess, and it became a Cold War instrument in addition when it turned out to be one of the few radio telescopes in the West to be able to track the rocket with which the first Sputnik was launched in 1957.[8]

The Changing Astronomical Community

Because of all these developments, the astronomical community of the 1960s was different from the community that founded the IAU half a century earlier. It was much bigger and more diverse—that is, more diverse in terms of research topics, mentality, and professional background, including physicists and engineers besides astronomers. The geographic diversity of the members also grew as more nations joined the IAU, albeit at a much slower rate; the Western countries remained dominant, with America providing about one quarter of all individual IAU members.[9] The gender diversity hardly changed, with the fraction of women well below 10%. There was no attention yet to these issues at the time: they were hardly ever discussed, except perhaps diversity in terms of seniority: the participation of young astronomers (see below).

---

[7] GA newspaper 1970, 14.
[8] Agar (1998).
[9] DeVorkin (1999), 98.

Joseph Pawsey, here seen at the IAU GA in Moscow, played an important role in bring-ing radio astronomy 'into' the IAU. Photo: Courtesy of the CSIRO Radio Astronomy Image Archive

The case of radio astronomy demonstrates that new (sub)disciplines were not automatically included in the Union. On the other hand, the astronomi-cal aspects of the new discipline *were* integrated. In this way, the boundaries of the discipline were kept clear: all astronomy was included, but nothing else. There were collaborations and joint Commissions with other Unions for boundary cases such as radio frequency protection or monitoring the Earth's rotation, but the IAU remained clearly the International *Astronomical* Union.

Something similar happened with solar physics, which tended to develop into a separate discipline in the 1960s, with its own journals. It stayed within the Union, just like astronomical space research (although the division of tasks between the IAU and COSPAR, the international Committee on Space Research, was contentious). That this was not always easy is illustrated by the fact that Pecker mentioned 'maintaining the unity of astronomy' as one of the core aims of the Union in 1967.[10]

---

[10] J.C. Pecker to EC, 22 September 1967, IAU Archives (old part) II.14.D.10; cf. Hufbauer (1991), Thomas (1999).

The General Assemblies were no longer enough to keep up, even though they lasted two full weeks: but they were only held every three years. The IAU could easily have lost its scientific relevance to other, more frequent conferences. This happened in several other Unions. But in the case of astronomy, the IAU kept cutting-edge scientific discussions 'on board' by establishing standard formats for smaller meetings: symposia and colloquia (they were later included in the scientific programme of General Assemblies in GA years).

Lively discussions at the 1975 IAU Symposium 72, in Lausanne: (Left to right) William Bidelman, Roger Bell. William Morgan, Philip Keenan and Bengt Strömgren. Photo: J. Andersen

Symposia were small-scale, workshop-like meetings on a specific topic, jointly organised by two or more IAU Commissions. Crucially, participation was by invitation only, a principle that was sometimes questioned or simply ignored, but never explicitly changed. Many were initially organised just before or after General Assemblies in the same country, to facilitate participation in both. The first symposium was a joint meeting with the International Union for Theoretical and Applied Mechanics (IUTAM) on Cosmic Aerodynamics in 1949, and several others were held in conjunction with the General Assembly in Rome in 1952. Still, the symposium on Galactic Research in Groningen, the Netherlands, in 1953 is known as Symposium no. 1, probably because it was the first to have its transactions published. General Secretary Pieter Oosterhoff then decided to make the symposia transactions a new series of IAU publications. Ever since, the Symposia have been known by their number, just like the Commissions. Their transactions, now approaching 350 volumes, are a testimony of front-line astronomical research in the past seventy years.

From the IAU Colloquium 6, held in Elsinore (Denmark), in 1969: (Left to Right) Jorge Sahade, Bohdan Paczynski, Daniel Lauterborn, Daniel Popper, Albert Linnell, Tibor Herzeg, Horst Mauder, Masatoshi Kitamura and Peter Conti. Several other senior astronomers are outside the photo. Photo: J. Andersen

Colloquia were less formal than symposia. The first one was organised in 1959 in Uccle, Belgium, on stellar photospheric structure. Colloquia were open, but more specialised, organised by one Commission, and their proceedings were mostly not published (at least not as a fixed series). All symposia and colloquia still had to be approved by the Executive Committee before they were allowed to use the IAU label. The IAU then also provided some funding, largely to assist attendees. In this way, the IAU could have some control over the quality of the meetings. This, in turn, enhanced the authority and prestige of the IAU label, which automatically became associated with high-quality scientific meetings, because of the keen competition for approval as IAU meetings. Equally important was that the new meeting formats gave Commissions the possibility to be more active between General Assemblies, and for researchers to become active in the Union. More than one future President or General Secretary started that way, as you can see in the interviews in this book.

Jorge Sahade and Kwan-Yu Chen (right, facing the camera), here seen at the IAU Colloquium 6 in 1969. To the left: Tamriko and Richard West; in the background Daniel Lauterborn. Photo: J. Andersen

Throughout the turbulent scientific and political developments of the Cold war decades, the IAU kept sponsoring its traditional astronomical services. A new service was the Information Bulletin on Variable Stars, started in 1961. It was published by László Detre of Konkoly Observatory in Hungary under the supervision of Commission 27. The much older Central Bureau for Astronomical Telegrams (CBAT) bureau in Copenhagen, which provided quick alerts on supernovae, comets, and other transient phenomena, moved to the Smithsonian Astrophysical Observatory in Cambridge, Massachusetts, in 1965. It was initially coordinated by Owen Gingerich, but was soon taken over by Brian Marsden.

In this period, the Minor Planet Center, still led by Paul Herget with a small staff in Cincinnati, not only processed new observations, but it also had a large job in verifying and merging the older lists of minor planets and looking for ones that had been observed before but had been lost since. As we saw above, it increasingly used computers to calculate the complex orbits. The MPC also maintained the list of names for newly discovered minor planets.

Time is timeless, so to speak, and the work of the Time Bureau would seem to be unchangeable—defining time was one of the oldest functions of astronomy, and accurate time was always needed. But time, too, was affected by the new technology. Better instruments made it possible to monitor the Earth's rotation with ever more precision. In 1956, different standards of time were recognised, with UT2 (astronomical time corrected for various factors, including seasonal variations in the Earth's rotation) being the most important, as it was used for radio time signals. But around the same time, atomic clocks provided a non-astronomical alternative for fundamental time. From 1961, the Bureau International de l'Heure (BIH) coordinated Coordinated Universal Time (UTC): atomic time, corrected to remain in sync with astronomical time (after 1972 this was done by introducing occasional leap seconds). This became the most widely used time. It was recognised by IAU resolution in 1967.[11] From 1965, the Bureau was directed by Bernard Guinot. The IAU's share in its funding diminished, as the maintenance of accurate time became increasingly the domain of physical metrology; the task was eventually transferred to the Bureau International des Poids et Mesures (BIPM) in 1988.[12]

Another task that was only visible for the specialist was the redefinition of the system of fundamental astronomical constants. In the 1970s, various Commission 4 working groups proposed a coherent new system of astronomical units, constants and reference frames that was adopted at the General Assembly in Grenoble in 1976. It has been updated several times since, most recently in 2018.

Teaching Astronomy

Defining time and reference systems, coordinating nomenclature and maintaining astronomical message services were all traditional functions of the IAU, just as the General Assemblies, symposia and colloquia were extensions of its core function of organising international scientific meetings. But gradually, the IAU expanded the scope of its activities. We have seen the beginning of this with the, admittedly small-scale, exchange programmes just after the war. From the 1960s onwards, the Union more actively sought to engage young

---

[11] On time: McCarthy and Seidelman (2009); the time-keeping functions of the BIH were taken over by the BIPM and the geophysical aspects by the IERS from 1988.
[12] Blaauw (1994) 266.

astronomers and researchers from developing countries—in general, a wider audience than the group of senior scholars who traditionally formed the IAU community.

Marcel Minnaert (right) is here seen at a Symposium in Utrecht in 1963. Looking up at Minnaert is Evry Schatzman. Also seen in the photo are Karl-Otto Kiepenheuer, Jaques Sauval and Robert Grant Athay (in front). Photo: H. Nieuwenhuijzen

At the General Assembly in Hamburg in 1964, a group of active astronomers, including Marcel Minnaert, organised a special session on astronomy education. At their initiative, Commission 46 on Teaching of Astronomy was founded, with Evry Schatzman and Edith Müller as its first Presidents. At the meetings, its members shared experiences and best practices. A special feature of this Commission was the national representatives or 'liaisons' in all countries, who could be astronomers with a special interest or experience in the subject, which in practice mostly involved astronomy teaching in secondary schools. One of the main activities of the commission was compiling a catalogue of the available educational material in various languages. It also started a project to collect and distribute educational material teaching materials (The Contretype Project). Both the catalogue and the contretype project quickly became too large and expensive for the IAU to handle, and as other projects proved more efficient, they were ended. The other activities of the commission continued, however, and other initiatives were added.

One programme especially turned out to work extremely well: the Summer Schools for Young Astronomers, later known as the International Schools for Young Astronomers (ISYA), initiated by Josip Kleczek.[13] The idea was first discussed at a joint meeting with UNESCO in Nice in 1964. The schools were especially intended for talented students from developing nations. The first ones were held in Manchester (UK) and Arcetri (Italy), but the next ISYAs were organised in Hyderabad (India), Cordoba (Argentina) and Lembang (Indonesia), with guest lecturers and researchers from all over the world (see list in Appendix D).

The objective of an ISYA was to broaden the participants' perspective on astronomy through lectures from an international faculty on selected astronomy topics, seminars, practical exercises and observations, and exchange of experience. ISYAs are scheduled in countries and universities with a long-term interest in the natural sciences, and to sustain further development of astronomical activities. In other words: they not only helped the students to start a research career, but they could also support the host observatory to develop its research programme.

In organising the schools in Manchester and Arcetri, young astronomers from various countries were brought to the observatory to work with local lecturers and instruments. At Hyderabad, a new scheme was tried: to bring the foreign lecturers and experienced research workers to an existing Observatory. This became the standard model for future ISYAs.

After this promising beginning, the ISYAs became a regular activity of Commission 46. Initially, the Schools lasted for 6–8 weeks during the summer. They were primarily sponsored by UNESCO as well as the IAU, while the host observatory provided housing and facilities for the 10–20 participants and the teachers. When UNESCO funding ceased in 1971, the IAU Executive Committee decided nevertheless to fund at least one ISYA during the triennium 1970–1973. The Schools thus continued with IAU support alone, but as the available funding was much less than previously obtained from UNESCO, the duration of the Schools was reduced by half, to about 3 weeks; still, the number of participants was sometimes over 50. Later, the ISYAs again received partial financial support from the UNESCO via ICSU, and the programme picked up, still primarily in developing countries and still greatly in demand, but now at a more realistic ambition level of about three weeks' duration. This has been found to be the minimum time needed for the

---

[13] Gerbaldi (2007).

participants to become accustomed to speaking and debating, in English and in public.

The ideal participants are students with a good college training in mathematics and physics (but not necessarily astronomy), usually from the geographical region of the host country. The lecturers are recognised experts from all over the world, who are asked to stay as long as possible in order for the participants to feel at ease in communicating with them. The curriculum includes lectures that cover the basic concepts and theories of selected fields of astronomy and astrophysics, along with relevant hands-on training in instrumentation and observation, data reduction and calibration, plus selected thematic topics of the latest developments and trends in the fields. Since their start, the ISYAs have proved enormously important for ca. 1400 young scholars from all over the world.

The exchange programmes and the International Schools for Young Astronomers were signs that supporting young researchers also became a task for the IAU. This was a new interpretation of its official aims (facilitating international co-operation and promoting the study of astronomy). The ISYAs also demonstrated that the IAU gradually started to look beyond the traditional strongholds of astronomy in Europe and North America.

Growth Pains

The triennial General Assemblies remained the most important meetings of the Union. The organisers claimed that the opening ceremony in Berkeley was attended by more than 10,000 people![14] This is where all Commissions met, resolutions were adopted, and new Executive Committee and Commission leadership elected. More importantly, high-level individuals from the global astronomical community attended the Assemblies, making them the most important networking events in astronomy. The informal contacts between sessions were at least as important as the formal talks, if not more. But this aspect came under pressure, or so many feared, by the changing scale of the meetings. Within a few decades, the General Assemblies grew into massive conferences, with more than 1100 participants in Hamburg in 1964, and more than double that in Brighton in 1970.

---

[14] GA newspaper 1961, 3.

Participants at the IAU General Assembly in Berkeley in 1961. Photo: IAU Archive

The conference always took two weeks, with a day in between for social events. It consisted of commission meetings, some ceremonies, and of course the General Assembly itself (the 'Business Meeting'), the meeting in which administrative issues were discussed and resolutions were approved. New elements were added to the programme after the Second World War, starting with Joint Meetings or Discussions (the terminology varied) of several Commissions on topics of mutual interest. Another new feature were invited discourses by prominent researchers, originally intended as public outreach lectures (Dublin 1955). Apart from the official programme, there were usually also excursions to nearby observatories, some concerts by local musicians (sometimes very special ones: Lady Susi Jeans, the widow of the astronomer James Jeans, and the famous radio astronomer Sir Bernard Lovell played the organ in Brighton in 1970), and other cultural events. There was also a Ladies Committee which organised activities for 'astronomer's wives', a sign of the gender balance in this period. In Berkeley in 1961, it arranged for morning coffees at various astronomers' houses, for example [sic!].

The growth of the General Assemblies could be interpreted as a sign of their importance and appeal, but not everybody was happy about it. Apart from the logistical challenges, the question was: could such a meeting, with so many different sessions, still be efficient? It became practically impossible to maintain an overview of what was going on in neighbouring (sub)fields. Besides, would it be possible to maintain the informal spirit of cooperation? At the opening of the General Assembly in Berkeley, IAU President Jan Oort, who had attended all previous General Assemblies since Rome 1922(!), confessed feeling uneasy about this: 'We must not run the risk that by an unwieldy size of our meetings the possibility of forming close ties and friendships is lost, or that some of our best astronomers refrain from attending because of disappointment about the too large meetings.'[15]

A common complaint was that the administrative meetings of the Commissions, in which for example their presidents and secretaries were appointed, took too much time, while they were only interesting for a small number of active members.[16] But the scientific sessions were also criticised. Nancy Houk, a self-professed 'young astronomer', complained that they had too many papers presenting highly specialised results, with too little time for discussion afterwards.[17] Should those kinds of talks not be reserved for smaller, more specialised meetings, reserving the General Assembly for more general discussions about the state of the field?

The most vocal critics of the General Assembly format included former and active Presidents and General Secretaries, such as Sadler, Goldberg, Pecker, Lubos Perek, and later Derek McNally and Johannes Andersen. Still, little changed. When a Committee on Future Organization, chaired by Pecker, sent out a questionnaire in 1962, it only got 10 reactions. This could either be a sign that the discontent was not shared widely—or that most members were simply not interested in these matters. Pecker's committee suggested having fewer and smaller Assemblies, with more specialised meetings (such as symposia) in between.[18] This did not happen, however.

Similarly, various attempts to organise special activities for junior researchers failed, leading Pecker's committee to conclude that it was a lost cause (others noted that the invitation was only circulated through IAU channels, missing the younger generation completely[19]). In any case, the Executive Committee agreed: 'The problem of young astronomers is largely exaggerated.'[20]

---

[15] GA newspaper 1961, 19.

[16] The IAU archives contain a lot of material on this issue; see especially box 6 and box 14. Notable critics included Sadler, Goldberg, Pecker, Perek, Lequeux, and later McNally and Andersen.

[17] GA newspaper 1970, 12.

[18] J.C. Pecker to EC, 22 September 1967, IAU Archives (old part) II.14.D.10; cf. proposals by D. Sadler, 13 May 1964, IAU Archives box 6, file 'Format General Assemblies 1964–89'.

[19] Hack, Marsden and Schatzman in GA newspaper 1970, 17.

[20] IAU Archives box 1A, Officer's meeting 27 October 1970.

Most of the scientific activity took place in the many Commissions. Here, too, adapting to the new scale and pace of post-war astronomy proved difficult. Otto Struve, director of the Yerkes Observatory, wondered already in 1946 if this might be a good moment to start afresh, with all new Commissions. This would also offer an opportunity to get rid of inactive members and even entire inactive Commissions. Interestingly, Harlow Shapley objected to these radical ideas on the ground that some 'dead wood' on the committees 'is part of the diplomatic interchange'.[21] These ideas represented contrasting views on what the Commissions were supposed to be: small working groups with specific tasks, or scientific bodies in their own right, consisting of all researchers working on the same topics. Struve's proposals were not adopted, but he had clearly touched upon a sensitive issue, and discussions about the functioning of the Commissions would resurface again almost every decade, well into the twenty-first century.

In the 1950s and 1960s, the Executive Committees tried to control the growth of the Union. In 1948 it was stipulated that members of the national committees of IAU member states would not automatically be members of the Union itself, in order 'not to unduly increase the membership of the Union'.[22] In 1957, 'general membership' of the IAU was introduced: astronomers could be member of the Union without belonging to any commission. The idea was that the Union could keep growing, while Commissions could remain small. This also did not work, however, so in 1961 another change was introduced: all Commissions got an organising committee, with a president and a vice president. The idea was to make it possible to have larger Commissions, without the risk that inactive members would slow down the work. The idea of Commissions as scientific communities had prevailed. The Commissions remained selective, however. All members were supposed to have 'the ability to contribute' to research on the topic, and they had to be confirmed by the Executive Committee and the General Assembly.[23]

At the General Assembly in Moscow, Pecker and Schatzman raised another problem with Commissions: some were organised by topic, others by the technology they used (for example Radio Astronomy), causing inconsistencies and overlap. This, too, would remain a topic of debate for decades, without major changes. Resistance against merging Commissions was always strong, for example. Apparently, the leading members (and perhaps would-be commission Presidents) were quite attached to their Commissions. In 1965, Sadler

---

[21] Blaauw (1994) 138; cf. idem, 209–219.
[22] Quoted in Blaauw (1994) 172.
[23] The debates are described in more detail by Blaauw (1994) 169–175; see also IAU Archives box 6, file 'Format General Assemblies 1964–89'.

proposed a compromise: grouping Commissions together in thematic 'Sections'. This too, was not implemented, although the idea was revived in the 1990s (see Chap. 5).

In 1967, the Statutes, By-Laws and Working Rules were revised, though no big changes were made; most changes involved bringing the rules up to date, and to clarify the procedures for new member nations and individual members. Interestingly, the rules also specified the required scientific merits of new individual members:

8. Individuals proposed for Union membership should, as a rule, be chosen from astronomers and scientists, whose activity is closely linked with astronomy taking into account

   (a) the standard of their scientific achievement
   (b) the extent to which their scientific activity involves research in astronomy
   (c) their desire to assist in the fulfilment of the aims of the Union.

9. Young astronomers should be considered eligible for membership after they have shown their capability (as a rule Ph.D. or equivalent) of and experience (some years of successful activity) in conducting original research.
10. For full-time professional astronomers the achievement in astronomy may consist either of original research or of substantial contributions to major observational programmes.
11. Others are eligible for membership only if they are making original contributions closely linked with astronomical research.[24]

In retrospect, it may not be surprising that most members were not interested in structural reforms during one of the most exciting decades in astronomical history. The science itself was far more interesting, and the General Assemblies have survived—in slightly modernised form—to this day. The rapid growth of the science also meant that many researchers were relatively young, focussing on building their careers rather than reforming institutions. At the same time, inequality or diversity still were not high on the agenda yet, as any problems concerned so few people.

The discussions about the size of the meetings and Commissions illustrate a fundamental characteristic of the IAU in its first half century: it was not designed to be an inclusive organisation. Expanding its reach or membership

---

[24] Working Rules, Proceedings of the 1970 GA 279.

was not a goal. One could say that the IAU *represented* the international astronomical community, but it did not *include* all of it. There was more worry about Commissions and meetings becoming too large than too small.

In other areas too, the IAU tried to prevent uncontrolled expansion. As mentioned, symposia and colloquia organised under the auspices of the IAU had to be approved by the Executive Committee, for example, in part to limit their total cost to the Union but also to ensure scientific quality (and enhance the authority and prestige of the IAU label). On a different issue, General Secretary Sadler commented in 1959 that he wanted Commissions to be more restrained with proposing resolutions, to prevent futile discussions in the General Assembly that might undermine the authority of the IAU.

Looking back in 2003, three General Secretaries who presented a systematic revision of the IAU statutes and by-laws stated that the 'underlying philosophy [of the old rules] was to control rather than to promote action'.[25] That was a hard but fair assessment. A lot of time and ink was devoted to controlling membership, both of the Union and of Commissions. The growth of astronomy, and of the Union with it, could never be controlled, however. Elaborate admission procedures were put in place. Yet in practice, only a small percentage of nominated candidates were turned down. All procedures were explained in an 'Astronomer's Handbook' from 1967, which was regularly updated. Individuals would only be a member of the Union through their Commission membership.

As the IAU and the number of its activities grew, the workload of the Executive Committee increased accordingly. Before the Second World War, the General Secretary could keep the Union running while corresponding with the Executive Committee (President and the several Vice Presidents), meeting them in person only during the General Assemblies. This was not enough anymore. After the war, it was quickly decided that the Executive Committee would meet annually. In 1951, the post of Assistant General Secretary was added—in practice a preparation for the Secretariat, since the AGS would automatically become the next GS. A few years later, the number of Vice Presidents was increased from 5 to 6. To keep the members informed of the increasing IAU activities, the General Secretaries also started to publish semi-annual Information Bulletins, from 1959.

All officers were elected for three years during the General Assemblies. After 1945, Presidents and General Secretaries usually only served one term, although General Secretaries also spent a term as Assistant General Secretary. The exceptions were Pieter Oosterhoff and Donald Sadler, who each served two terms as General Secretary between 1952 and 1964. In 1958, a separate

---

[25] Johannes Andersen, Hans Rickman, and Oddbjørn Engvold in IB 94, 32.

Special Nominating Committee was created to find candidates to be proposed for the next Executive Committee.

Donald Sadler here seen at the Hamburg GA. Photo: Courtesy of the Hamburger Sternwarte, Universität Hamborg/W. Dieckvoss

The members of the Executive Committee were all active astronomers. The General Secretary generally could rely on some secretarial assistance at his institute, but as the administration increased (think of the address lists of more than a thousand members!), full-time professional assistance became necessary. Nel Splinter was the first permanent executive secretary 1958, initially working in Leiden (Netherlands) with Oosterhoff, and then in Herstmonceux (Britain) with General Secretary Sadler. In 1962 Splinter was succeeded by Dorothy Bell. Both were highly respected, earning the titles of the first and second 'Miss IAU'.[26] They were succeeded by Arnost Jappel, the first male 'Miss IAU', a qualified lawyer from Czechoslovakia who was prevented from working in his field under the Communist regime.[27] The secretariat (read: the executive secretary, bringing the archive with her/him) moved with the General Secretary. General Secretary Patrick Wayman described them as 'peripatetic executive secretaries'.[28] Jappel worked in Paris with Jean-

---

[26] Blaauw (1994) 208, 233.
[27] Pecker, 'Memories of IAU', GA newspaper 2006, 2.
[28] GA newspaper 1979, 11.

Claude Pecker, then with Perek in Prague and de Jager in the Netherlands, and finally with George Contopoulos in Thessaloniki and Athens. The permanent office in Paris was only established in 1979.

One of the most important tasks of the secretariat was organising the General Assemblies, together with the local organising committees. The latter also consisted of active astronomers and students. As the General Assemblies grew, they became harder to plan and manage. The General Assembly in Brighton in 1970 was the first one that involved professional conference organisers. General Secretary Perek also warned the Executive Committee of another modernisation: 'Dear Colleague, Although none of us is very enthusiastic about newspapermen, some degree of publicity is unavoidable'.[29] The IAU increasingly had to represent astronomy to the outside world as well as to the astronomical community itself.

IAU General Secretary Kees de Jager giving an interview at the 1973 GA in Sydney. Photo: Courtesy of the CSIRO Radio Astronomy Image Archive

---

[29] L. Perek to EC, 6 March 1970, IAU Archives (old part) I.14b.F.

Jean-Claude Pecker, Astronomer, Painter, Poet … and Mr IAU

Assistant General Secretary (1961–1964), General Secretary (1964–1967)
   Interview conducted in Paris on 1 March 2017
   Few scientists have been more prominent in working with the IAU than French astronomer Jean-Claude Pecker, the first officially-appointed Assistant General Secretary of the IAU (1961–1964) and, as has now become the rule, General Secretary (1964–1967) and finally Adviser to the Executive Committee (1967–1970). But obviously his encounter with the IAU began much earlier.

Jean-Claude Pecker, at 93 years of age, in his modest apartment in Paris' Quartier Latin district. Photo: C. Madsen

It wasn't obvious that Pecker would become an astronomer. Born in Reims in 1923 into a Jewish family that hailed from Alsace, young Jean-Claude, with his parents (his father being an electrical engineer and his mother a teacher of philosophy and literature), moved to Bordeaux, where they lived in 1940 when Germany unleashed its *Blitzkrieg* and took control of France. Perhaps surprisingly, and not entirely voluntarily, the family moved to Nazi-occupied Paris, where his father could work in a less exposed position, in a 'back-office function', thus considered somewhat safer. In Paris, young Pecker applied to join the prestigious École Normale Supérieure. Though he was accepted, as a Jew in those dark days joining was in the end not a feasible option. In fact, on his 21st birthday, his parents were arrested, deported to Auschwitz and subsequently murdered. Himself avoiding arrest by assuming a false identity and working for a while in a Paris factory, Pecker survived. He even published a paper during that time under the assumed name of Jean-Claude Pradel on electrolytic polishing in a popular science magazine.

'After the liberation of Paris, in September 1944, I began my studies at the Physics Department of the ENS, with Alfred Kastler (1962 Nobel Prize in Physics) as my academic mentor.'

(Continued)

For his PhD studies, Kastler advised Pecker to go into theoretical astrophysics, a field that was not yet covered in France but developing rapidly. Pecker chose stellar atmospheres as his field of study.

With Evry Schatzman, a friend and intellectual mentor, he published his first astronomy paper in 1947 on the determination of the hydrogen and helium content of stars using tables of luminosity, mass and radius for about 50 stars.

'We got abundances for these stars, yet they were all wrong because we assumed that they were main sequence stars. But they were not! But anyway, I learnt a lot. In the meantime, Vladimir Kourganoff, who had worked with Svein Rosseland in Oslo, had translated Unsöld's book "Physik der Sternatmosphären". It was very useful. I really learnt much from that book.'

Yet to move on, it was necessary to work with fellow scientists abroad. 'But in 1947, it was difficult to travel in Europe because of currency restrictions and also housing problems. For a young man as I was then, working with other scientists elsewhere was simply not possible. No money, no house! At that time, Colonel Stratton was General Secretary of the IAU and the Executive Committee took the decision to create Commission 38 for the exchange of astronomers. It was later abolished, but at that time it began to organise the exchange of researchers from country to country, so we swapped money and housing between ourselves. This way, I could go to Utrecht in the Netherlands in exchange for Jacob Houtgast, who came to Paris to live in my house. In Utrecht, I stayed in Houtgast's place. Cornelis ("Kees") de Jager lived there, too, so we became very good friends. It was a splendid year working with Marcel Minnaert, who was director of the Observatory. So the IAU helped me to go and spend the best time of my scientific life there. The following year, I went to Copenhagen to work with Bengt Strömgren. This was also a magnificent year. In this sense, I consider the IAU as the Godfather—or the Godmother, if you will—of my scientific life.'

In February 1952, Pecker participated in an expedition to Khartoum to observe the Solar Eclipse of that year. In the same year, the VIIIth IAU General Assembly took place in Rome, where he gave a talk on the eclipse to Commission 13.

'At Castel Gandolfo, we shook hands with the Pope. At the Rome meeting Donald Menzel, who chaired Commission 13, subsequently invited my to join him at the High Altitude Observatory in Boulder, Co., as a Fulbright Scholar, which I did during the following year. From the Rome General Assembly, I have fond memories of a long discussion under the olive trees about the Universe with three people, who played a great part in my life—Fred Hoyle, Schatzman and Dick Thomas, who worked on stellar chromospheres.'

At the Xth IAU General Assembly in Moscow in 1958, Pecker and Schatzman took advantage of the opportunity offered by the GA 'newspaper' 'Kosmos', to publish an article proposing a revision of the IAU Committee structure, which they found to be outdated. Their argument was what they saw as a disconnect between committees dealing with technology and those dealing with science. Their ideas were subsequently discussed with the IAU Executive, at the time comprising Donald Sadler, Pieter Oosterhoff, and Jan Oort. In any event, the enthusiasm and engagement of Pecker in IAU affairs had been noticed by the IAU Executive Committee.

'Arriving in Berkeley for the XIst General Assembly in 1961, I found a note from Sadler in my pigeon hole, asking me to see him as soon as possible. I had no

(Continued)

idea what this was about. Sadler asked me to sit down, and then he asked me if I wanted to become "the next general secretary". He gave me a week to think about it—a time that I used to consult with a few friends and colleagues. Only one advised me to reject the offer with the argument that it would mean the death of my scientific career. But I decided that my scientific career would survive and so, I accepted'.

As Assistant General Secretary, Pecker was charged with organising the IAU publications, including the IAU Handbook. This was also the time when the IAU began to publish the collection of proceedings of symposia and colloquia, using a set format and a single chosen publishing company.

In 1964, the XIIth General Assembly was held in Hamburg.

'Hamburg was difficult for me, given my family background. The idea of having to shake hands with people who might have been Nazi officials was painful. But I was General Secretary. And I was not against the idea of having the meeting in Hamburg. In fact, that was good—I'm for a united Europe; it was just a personal thing.'

Before the Hamburg meeting, from his office in Liège, Pol Swings, who was set to become IAU President at the Hamburg General Assembly, called Otto Heckmann, informing him that the Special Nominating Committee suggested to nominate him to follow Swings as IAU President (effective from the Prague General Assembly) and asking for his acceptance. Pecker was present, too.

Jean Claude Pecker here photographed during the IAU Executive Committee meeting in, Liège. Photo: Leo Goldberg, courtesy of AIP Emilio Segrè Visual Archives

'Heckmann, who was a very honest person, hesitated, remarking that he had been a member of the N.S.D.A.P. I asked for the phone and said "Prof. Heckmann, I was one of the people recommending you for the presidency and you should know that my parents died in Germany during the War. So, Prof. Heckmann, please accept." He did.'

Reaching out to heal the wounds of the world war was not solely a question for governments and institutions. It was very much a personal affair and using the bonds of science helped.

'After the Hamburg General Assembly, as secretary General, my job was to look after "everything else" [than the publications]' as he modestly expressed it—organising the Executive Committee meetings, taking decisions about new commissions and symposia and any other matters. It was a demanding time.

(Continued)

'In 1961, I became Assistant General Secretary of the IAU. In 1962, I became Director of the Nice Observatory, run down at the time. In 1963, I was elected as Professor at the Collège de France, and in 1964 General Secretary of the IAU, but I received a lot of help from the "Misses" IAU, Dorothy Bell and Geneviève Drouin, and later from Arnost Jappel, who took over the running of the secretariat. Also, I had an excellent cooperation with my friend Luboš Perek, who succeeded me in my IAU functions as well as with my predecessor, Donald Sadler.

Evry Schatzman was very keen on the subject of teaching astrophysics, a domain that at the time was not well covered by universities in France. At the IAU he chaired the Commission on the Teaching of Astronomy from 1964 to 1967. Under his influence, I proposed the School for Young astronomers to the Executive Committee. We needed advice and help from UNESCO—and money! So I organised a small meeting in Nice attended by Michael Fournier d'Albe, then in charge of science teaching at UNESCO, and—from the IAU—Henk van de Hulst, Eugene Karadze, Vladimir Kourganoff, Schatzman and myself. The meeting concluded that there should be an International School for Young Astronomers. The IAU Executive Committee gave enthusiastic support to the idea, and UNESCO agreed to help finance it. The first school took place at Manchester University in 1967.

I ended my term at the Prague General Assembly in 1967. I really enjoyed Prague, dancing the night away at the closing meeting.'

Participants dancing at the GA Banquet in Prague. Photo: IAU Archive

Nevertheless, it had been 'heavy lifting', as one would say today, inevitably taking its toll on Pecker.

(Continued)

At the same time, change was in the air in Europe. A few months later, student unrest broke out in Paris, while the so-called Czech Spring with calls for democracy enveloped Prague and the Czech nation and challenged the Soviet supremacy over Eastern Europe. The incoming General Secretary, Luboš Perek, would see himself confronted with the unfolding developments in his home country. Pecker, the Parisian, who suffered from exhaustion finally found time to pursue his own interests travelling to the Americas and Australasia. Science aside, it included painting watercolours and writing poetry.

The IAU, however, would also benefit from the artistic skills of Pecker. In the 1970s, the question came up about a 'symbol' for the IAU—what today would have been described as a 'logo'.

'Actually, the question was in the air since the tenure of Donald Sadler, when he was General Secretary and I was Assistant General Secretary. We had discussed the matter at length. In Sydney, my old friend Kees de Jager ended his term General Secretary but continued as an advisor to the Executive Committee, whilst I had become chairman of the IAU Resolution Committee; hence I was still very close to the Executive Committee. On request, I made a few sketches for a symbol—all bilingual of course—perhaps half a dozen, which were submitted to the Executive Committee. After minor suggestions, one of my proposals was accepted and is still in use.'

Even though he served a term as an advisor to the Executive Committee after his time as General Secretary, thus following the established practice of the IAU, Pecker 'returned' to science, moving from his original interest in stellar atmospheres to the study of Cepheids and ultimately to cosmology. Today, the history of science is the focus of his work. Yet he continued to follow IAU affairs—indeed also attending every General Assembly since then, save for the 2016 one in Honolulu. 'I had bought the ticket, but in the end, health issues prevented me from travelling. Now, I hope I can make it to Vienna.'

In the end, he did not, but he did make an impressive presentation there by means of video. And so, always engaged and outspoken at the General Assemblies, at least for some, he has become 'Mr IAU', though he himself would point to other scientists as candidates for this title.

Luboš Perek, Astronomy Under the Cloud of Totalitarianism

Assistant General Secretary (1964–1967), General Secretary (1967–1970), President of Commission 33 Structure & Dynamics of the Galactic System (1973–1976)

Interview conducted in Prague on 31 October 2016

For many young astronomers in Post-War Europe, the IAU was a door-opener to the international community by supporting cross-border travel and work. For those who found themselves behind the Iron Curtain, the situation was especially dire and the help by IAU, limited as it may have been in the face of the overall problems, was even more important.

(Continued)

Luboš Perek in his apartment at Kourimska 28 in Prague. Photo: C. Madsen

'For me, the most important features of the IAU have been the opportunity of personal contacts and friendships, and the fact that the IAU has taken care of young astronomers' needs, including travel grants either directly or via UNESCO. For me, the IAU meant the first meeting with "real astronomers".'

Born in 1919, the year the IAU was established, Luboš Perek was at an age where many would open the most exciting chapter of their lives as university students, when in 1938/39 Nazi-Germany occupied and annexed Czechoslovakia. The universities were shut down and remained closed until the end of the war. Travelling too was impossible. After the war, the daunting task of rising from the ashes awaited a devastated continent.

'In 1946 Prof. Josef Mohr invited me to come to the Masaryk University in Brno as an assistant. At the time the entire possessions of the institute comprised a crate of books and a piano. There was no astronomical telescope in Czechoslovakia—a 60-cm telescope had been taken over by Hungary as it was located in the part of Eastern Slovakia that was ceded to that country following a brief military conflict in 1939 (the telescope was later moved to the Tatra mountains). Then in 1948, the Communists took over in Czechoslovakia.'

'Thankfully, Mohr protected the faculty from too much meddling from the Communist Party. This meant that students who were not communists could study. But we were being watched. And one day Mohr received a letter from the party saying "throw Perek out!", but he kept the letter in his drawer.'

At that time, Perek was in Leiden on a UNESCO grant. 'I couldn't take my wife with me... it was not permitted,' he remembers.

While staying in the Netherlands and sharing an office with Adriaan Blaauw, he obtained a copy of the blueprints for the Dutch 50-cm Zunderman Telescope. He subsequently brought the drawings back to Brno, where—taking advantage of an existing piece of glass—the primary mirror for the new Czechoslovak telescope could be increased in size to 60 cm.

'In 1961 Bohumil Sternberk became one of the IAU Vice-Presidents and he put my name forward for the job as General Secretary, which by that time meant first spending a three-year term as Assistant General Secretary. It was agreed, so in 1964 I started to take care of the symposia.'

(Continued)

Pol Swings and Bohumil Sternberk at the GA in Prague in 1967. Photo: IAU Archives

With the departure of Dorothy Bell from the IAU secretariat, Perek suggested Arnost Jappel, a compatriot of his, as the new secretary. Jappel was a qualified lawyer and mastered many languages. However, not a supporter of Communism, he couldn't work in his field and mainly lived as a translator. Jappel accepted the offer and left for Paris to work with Jean-Claude Pecker. From 1964, he worked with Perek in Prague, then with Kees de Jager in the Netherlands and finally with George Contopoulos in Thessaloniki until 1979.

Reading the GA newspaper '*Nvncio Sidereo*' at GA. Photo: Masaryk Institute and Archives of the CAS

(Continued)

'Then in 1967, at the General Assembly in Prague, I took over as General Secretary, followed by a term as advisor to the Executive Committee until 1973. As General Secretary, together with Donald Sadler, who was Director of the Nautical Almanac Office, I organised the 1970 General Assembly in Brighton.'

Perek realised that with the growing number of participants to the General Assemblies, the involvement of professional conference organisers was becoming necessary, and thus the Brighton meeting was the first one that availed itself of this opportunity.

'Also, in 1967 the IAU faced financial problems. Expenses exceeded income and the banking accounts showed a fifty per cent decrease. Increasing national contributions of some countries would have changed the balance within the Union, not a desirable situation. To overcome this, the Executive Committee decided to raise the "unit of contribution" by 50%. An increase of the unit thus did not change the balance between countries. It was also necessary to stop the free distribution in the IAU Transactions and to find a less expensive publisher. These measures permitted the IAU to retain the important principle of individual membership, rather an exception than rule among scientific unions.'

The IAU enabled Perek to maintain good connections with the international astronomical community, but the life in a totalitarian system remained a constant challenge. Perek himself chose a diplomatic description of the often precarious situation: 'You could make waves, but not too big waves.' In 1968, during the period known as the Czech Spring, the pressure eased somewhat. '"The Spring" really started in 1967, and professionally it meant fewer problems. It also became easier to have international visitors.'

But it was a brief respite. 'To meet with Otto Heckmann and the IAU Executive Committee at the ESO offices in Hamburg in August 1968, I took the car together with my wife Vlasta and Arnost Jappel. We first travelled to a hotel near the German border. To our surprise we learned in the morning that the border had been closed because of the invasion of the Soviet army. Yet we found the border open and no soldiers on the horizon. But in the first German village, there was a group of Czech people all listening to Radio Prague and talking about the Russian occupation. Upon our return, we found the border open, but a female guard was crying and two Russian soldiers were crouching behind a tree. It was a sad return.'

That same year Perek became director of the Institute of Astronomy of the Czechoslovak Academy of Sciences. In the municipality of Ondřejov, where the observatory was located, 'the party' was very strong and the local cadres moved to have all the activities and staff of the institute brought there. The local Prague party resisted and was, in any case, more 'reasonable', Perek found. To support their fight, they urged him to join, and he finally gave in hoping that the spirit of the Czech Spring could be retained. Alas, this hope was in vain.

'I served as General Secretary under President Otto Heckmann. He was a real gentleman and he wanted to do something for me. He invited me to visit La Silla in March 1969, at the time of the inauguration of La Silla. I was offered two weeks of observing time at the 1-metre photoelectric telescope equipped with an integrating photometer. I measured the H-beta brightness of 117 nebulae in the Southern Milky way. Thus I came to La Serena, visited briefly the famous bay of Coquimbo, before travelling to La Silla. I could use a car for travel on the mountain. In return, I brought with me a copy of the Catalogue of Planetary

(Continued)

Nebulae by Luboš Kohoutek and myself, which I presented to the library on the mountain. While at La Silla I met with my old office mate Adriaan Blaauw, and other friends, and even had the opportunity to talk with Olof Palme, before travelling by car with Heckmann back to Santiago.'

As Minister of Education Olof Palme attended the official inauguration of ESO's La Silla Observatory, representing the Government of Sweden.

In 1975, he became appointed director of the United Nations Outer Space Office (UNOOSA), then based in New York. He stayed in New York until 1980, when he definitively fell into disgrace of the Communists in power and to protect his family back in Czechoslovakia he was forced to return to his native country. However, for a while he managed to follow the activities of the International Astronautical Federation, of which he subsequently became President. Perek also chaired IAU Commission 33 (Structure and Dynamics of the Galactic System) after his time in the USA. 'But after that they let me feel their heavy hand. All the actions which I had discussed with people for many years—everything went down.'

Luboš Perek addressing the audience at the opening of the 2006 GA in Prague in 2006. Photo: IAU Archives

But the world keeps changing. Thus in 2006, the IAU returned to Prague for its XXVIth General Assembly. Thus Perek once again found himself involved in the IAU as a co-organiser of this meeting working with Jan Palouš and others. It was a time when the Czech Republic was becoming fully integrated into an open and free Europe, both politically, culturally and scientifically, including a successful effort to join ESO in the following year. The struggle for a free science, unimpeded by political ideologies was over.

Michèle Gerbaldi, Passionate Educator

President of Commission 45 Stellar Classification (1997–2000), Chair of the International Schools for Young Astronomers (1994–2007)

Interview conducted in Paris on 18 May 2018

Astronomy teaching may not have been at the centre of attention for the young IAU. But by 1964 the topic had gained enough importance to warrant the

(Continued)

establishment of a dedicated Commission—No. 46. Since then, astronomy educa-
tion has developed to become one of the key activities of the Union, manifestly
demonstrated by the Strategic Plan 2020–2030. One of the persons who drove
this development forward with energy, dedication, and—yes—passion, is
Michèle Gerbaldi.

Michèle Gerbaldi in front of the historical building of Paris Observatory. Photo:
C. Madsen

'My affection for astronomy began very early in my life. On every clear night,
before going to bed, my grandfather showed me the stars. We had a garden
with many trees, and I placed my little box camera in one of the trees and took
long-exposure photos of star trails.

During my studies at the Université de Paris-Sud (Orsay) doing mathematics, I
met Vladimir Kourganoff who taught astronomy. He recruited me as an assistant
at the University, and I did my master in astrophysics in 1968—yes it was in Paris
and I was there! Then in 1969, I started on my Doctorat d'Etat (which no more
exists) at the Institut d'Astrophysique de Paris, a laboratory of the CNRS, working
on the atmospheric properties of the peculiar stars Ap and Am.

My first IAU General Assembly was the one in Brighton in 1970. It was fantastic. I particularly remember seeing some of the famous authors of the books we
had in our library, including Ellen Dorrit Hoffleit who had worked on the Yale
Bright Star Catalogue, which was so important to us, as an observer.

At Brighton I attended the meetings of the Commissions on stellar classifications and stellar atmospheres, but also the one on the teaching of astronomy,
because I was very interested in teaching.

One challenge back then was the need for illustrations for teaching purposes.
At the time, I had begun to bring back photographs from the Haute Provence
Observatory to use during my teaching and that became the humble start of my
contribution to the Contretype Project (CP)—a collection of astronomical images
copied onto 35-mm slides for education purposes and made widely available.

(Continued)

Discussions about this project began at Brighton, led by the Commission 46 (Teaching of Astronomy) President Edith Müller. At that meeting, Evgeni Kharadze proposed a project to establish a clearinghouse where astronomical pictures of current interest could be made available for reproduction to anyone interested for educational purposes.

At the suggestion by Edith Müller, Kharadze, Edvard Kononovich and Donat Wentzel were put in charge of the project and—upon the nomination by Evry Schatzman—I agreed to act as secretary of this group. We then contacted the major observatories asking them to supply pictures to the project.'

Of the several dozen institutions asked to contribute, seventeen responded. Most of the material originated from French observatories.

In 1973 at the IAU General Assembly in Sydney it was then decided to start a pilot scheme in Moscow and Paris, respectively, and to have the material reproduced in each place upon request. The Paris collection, the main one, consisted of 56 slides. 'In Paris we received orders for about 1800 slides to be reproduced. In the beginning, the orders were all placed by astronomers.

It was clear, however, that the aim of the Project to get good teaching materials into the hands of teachers, particularly in the developing countries had failed. In order to make this project known in developing countries I announced it in the Courrier de l'UNESCO as an IAU project funded by France and soon I received requests from all over the world.

As during the second half of the 1970s many observatories and planetaria had begun to produce slides by themselves, even in colour, the decision to stop the Contretype Project was taken at the Montreal General Assembly in 1979.

Meanwhile the project was even expanded in France with my colleagues, Lucette Bottinelli and Lucienne Gouguenheim at the Université de Paris-Sud by assembling two sets of 24 slides each, on the subjects of gravitation and of spectral analysis. Each set came with a booklet giving not only a description of the slides but an analysis from a physical point of view, useful for the secondary school teachers.'

Back at the General Assembly in Brighton in 1970, another project had also started. For the first time, with the collaboration of the Commission members from 22 countries, a worldwide list of Astronomy Educational Material (AEM) was prepared to comprise the available teaching material recommended by astronomy teachers of IAU member countries. It was intended to make it available to any astronomer or science teacher involved in the teaching of astronomy.

'The idea had been suggested by Commission 46 already during the Prague Assembly in 1967 and soon thereafter, the first worldwide list of the AEM appeared, not the least as a result of hard work undertaken by Edith Müller.

The AEM project was updated every three years at the time of each General Assembly, and the interest for such a list was very high.

There were three lists: Books in English, books in the Slavic languages, and one for books in all other languages. The lists were distributed to institutes across the world once more with financial help from France. At the General Assembly in Patras in 1982, I became involved in the publication of "the third" AEM list. This was boosted significantly when we received a substantial additional list from China—with no less than 424 references of publications since 1949. This happened when Richard West was General Secretary—and he agreed that this time,

(Continued)

the IAU should carry the not insignificant cost for the distribution. Aside from its own merits, it was obviously an important political signal of respect and welcome to China, shortly after its re-entry into the IAU.'

This notwithstanding, at the General Assembly in Buenos Aires (1991) this activity came to an end—it was no longer possible to keep track of all the publications that by that time seemed to mushroom almost everywhere.

'At the same time, I became Deputy Director for the International Schools for Young Astronomers, the ISYAs, with Donat Wentzel as the Director. The formal titles at the time were different, though, but I use the titles as they exist today.'

The first ISYA was organised in Manchester in 1967 with a second one in Arcetri in the following year. The idea was to help young astronomers, especially from developing countries, introducing them to modern astronomy. This would be done by means of a course of up to 8-weeks duration, initially organised by Josip Kleczek on behalf of the IAU but with funding from UNESCO.

'However, it soon became obvious that the courses should rather take place in the developing countries, and thus started a long series of ISYAs there with the first one in Hyderabad in India, in 1969. During "my time" we had ISYAs in China, India, Egypt, Brazil, Iran, Romania, Thailand, Argentina, Morocco, Mexico, and Malaysia.

The ISYAs had other objectives, no less important than the "pure" learning goals. We wanted to broaden the view of the students—allowing the students to retain a broad perspective on our science and not only stick to a single, highly specialised branch of research. That also enabled them to make a critical choice for their life: To pursue a research career or not.

We also wanted to combat the "lonely astronomer syndrome"—the fact that in many countries, astronomers found themselves professionally isolated, because they were the only ones in their place and often lacked the means to develop a personal network. In fact, for most of the students it was the first time that they were—and are—learning and working in an international environment, receiving lectures by astronomers with an international reputation and coming from all over the world.

Boarding and lodging are in the same place for the lecturers and the participants and the informal conversations strongly help the students to become more confident, to discuss their ideas, to ask questions in face-to-face situations and to start to build their international network. Experiencing the surroundings of the host country and the discussions that arise between participants and lecturers from different cultures are as equally important to build long-term perspective among the School alumni and the lecturers.

I believe this has been quite successful. For example, when we had the General Assembly in Kyoto in 1997, several of the former students came and we shared a pleasant evening together. Indeed the ISYAs have contributed to building an international community. So, it was never "simply" about teaching.

In the ISYAs, we brought together young people who all shared a passion for astronomy, but had very different backgrounds. I remember a School in Egypt in 1994, where we had students from the Middle East Region who were orthodox, Catholic, Muslim and atheists, and yet thoroughly enjoying the gathering. Also notable, when the schools took place in Malaysia and in China, we also had students from North Korea. For the participants these Schools were—and are—a

(Continued)

cross-border initiative that transcends the political and cultural barriers through their common interest: astronomy.

Building bridges between different cultures: ISYA in Iran 1997. Photo: Collection J. Andersen

A direct impact of the ISYA programme can be noticed when looking at the names of the ISYA organisers of a host country today. Quite often the organiser has been attending a previous ISYA as a student perhaps 10 years ago.

Regrettably, the UNESCO funding was stopped in 1971—although between 1979 and 1990 some UNESCO money was given to the programme through ICSU. In any event, with a reduced budget, both the number of schools and their duration had to be reduced. Luckily, the new partnership between the IAU and the Norwegian Academy of Science and Letters have enabled us to increase the activity again.

From a programmatic point of view, I think that the ISYAs can be divided into two epochs, the early phase and a second phase beginning in the 1990s. Back in the 1970s, the first computers for routine use in research—with punch cards— came about. Also, microphotometers became available to scan photographic plates and the ISYAs obviously became a vehicle for introducing these new techniques to the students. In the 1990s there was another big jump with the emergence of databases, not the least the CDS in Strasbourg. Now, with interlinked computers connected via a telephone line, we could begin to use the databases—also for teaching. Also, the IUE space experiment database played an important role in providing real astronomical data for teaching. Today it's routine, of course, but back then, it was a real paradigm change in astronomy teaching.

(Continued)

I "experienced" my first ISYA in China in 1992. There was no way to link to other computers, but I arrived with data tapes and that gave the students a first feeling for how to work with a (real) database.

A different activity that I'd like to mention was the TADs, the Teaching for Astronomy Development, implemented by Don Wentzel in 1994. The TADs had a more limited scope—meaning fewer participants, mostly only from the hosting country, and shorter courses. But they were less costly, of course, and so, the IAU could organise many TADs across the world. The first one was held in Vietnam in 1996 with others in various places in Central America, Morocco, Kenya, the Philippines, Mongolia, in the Caribbean and elsewhere. The idea was also that after holding a TAD, a country might later apply to the IAU for an ISYA. The last TAD took place in Kathmandu in 2008, before the programme was folded into the overall remit of the OAD in Cape Town. Having the first TAD in Vietnam had a strong symbolism after the 30-year hiatus from the international community.'

From a humble start in 1967, the education activities under the IAU auspices had grown in complexity with the ISYAs, the TADs and the IAU Travelling Telescope Project that mainly came to use in Latin America. They were all led by the enthusiasm of individual members, but also supported by the IAU Office.

'All the general secretaries gave their support to this endeavour. They did whatever they could to simplify the process—from communication and money transfer to helping with travel and possible immigration issues. It could not have worked without it, because you often need to take decisions on very short notice. Until recently, the ISYA programme was under the direct supervision of the IAU General Secretary and Officers. Of course, for me, in many ways helping to manage the ISYAs was simplified by the fact that I was located at the IAP, i.e. in the same place as the IAU office. So it was an easy way for me to report periodically to the General Secretary and Officers of the progress of the next ISYA.

And I can say with certainty: There was no bureaucracy at all. Even so, we faced other challenges. One was—and still is—the lack of recognition of the activity within the national academic systems. At university, you're supposed to do research and teach your students. Spending two months on an ISYA in a far-away country hardly counts—if at all—towards your own career objectives.

Also, up until 2008, inside Commission 46 discussions took place whether it was feasible for the IAU to include teaching activities involving school teachers. Be that as it may—with the adoption of the first IAU Strategic Plan in 2009 and the subsequent establishment of the OAD and the Office for Astronomy Outreach, the OAO, the IAU took a major step forward. It is true, however, that this has led what used to be Commission 46, now C1, in search of a meaningful role within the teaching portfolio. One new focus is to look at research in education, which can bring added value to the overall endeavour as well as activities of working groups with specific tasks, e.g. astrobiology.'

In 1997, Gerbaldi took over as leader of the ISYA programme until 2007, with Edward Guinan as deputy. She was succeeded by Jean-Pierre de Greve with Kam-Ching Leung as deputy. Yet she continued teaching at ISYAs for several more years.

'Astronomy has a special place among the sciences in being able to attract young people. Carrying out their own astronomical observations is a great experience for undergraduates. The observations they do, e.g. images, are their very

(Continued)

own observations. Astronomical images can seem magical. But they are not magical at all. They are very rational, very scientific.

Activities in support of astronomy teaching exist in the USA and in Europe—in Europe ranging from EU-supported activities to those of ESO, but they hardly happen in the rest of the world. Here, the IAU is in a very special situation. We go to these countries, we facilitate international exchange in areas that struggle with all kinds of obstacles and we contribute to the development of their societies. Astronomy teaching is not at all only about astronomy itself. The IAU is simply the only body that can reasonably undertake this task, partly as a truly international body, partly because it's recognised as a *bona fide* organisation by governments. If the IAU would not be doing it, it would not happen.'

Gerbaldi's passion for teaching has never ceased. She is now involved with OAD activities with a strong focus on African activities and novel teaching approaches. Indeed, the idealistic fire of the 1968 Orsay student, now augmented with a lifetime of hard-earned practical experience from organising courses and teaching across the world, has stayed with her.

'In my latest activity together with Jean-Pierre de Greve—in Southern Africa—we make the students part of their own teaching, to learn themselves by asking questions and by developing their ways of thinking.'

4

A Two-Faced Period (1970–1989)

Expansion and Stagnation

The 1970s and 1980s are a somewhat ambiguous period for the IAU, characterised by consolidation while waiting for the next great step. Astronomy kept expanding, mainly driven by instrument development. The succession of discoveries of the 1960s, based on the opening of the electromagnetic spectrum, was followed by more detailed and systematic observations in all wavelength ranges. Space-borne instruments became standard parts of the astronomical toolkit. Planetary missions brought the solar system into close view. A range of 4-metre class optical telescopes was opened, which became the workhorses for astronomy in the next decades. At the same time, planning and preparations for a new generation of 8–10-metre size telescopes were started, spearheaded by the experimental 4.5-m Multiple Mirror Telescope.

Many of the new instruments were developed by international consortia, some ad-hoc, some strongly institutionalised. Large instrument projects became the main drivers of international cooperation. The data that these instruments produced was exchanged more and faster, both the raw data, now increasingly digital, and publications in peer-reviewed journals, which all but replaced the older observatory publications and national journals.

Science flourished within the IAU too, especially at the Symposia and Colloquia. The International Schools for Young Astronomers (ISYAs) became a fixture in the astronomical community. Every year, a few dozen graduate students were introduced to front-line research by an international line-up of lecturers through the ISYAs. Especially students from developing countries benefited. Astronomers from countries with smaller astronomical communities

© Springer Nature Switzerland AG 2019
J. Andersen et al., *The International Astronomical Union*,
https://doi.org/10.1007/978-3-319-96965-7_4

also benefited from the new regional meetings that were introduced from the 1970s in Europe, Latin America and the Asian-Pacific region.

The Regional Meetings brought together astronomers from nearby countries, but sometimes they also included popular outreach activities. This photo shows the strong interest in David Malin's colour pictures of galaxies and nebulae at the 5th Asian-Pacific Regional IAU Meeting (APRIM) in Sydney. Photo: CSIRO Radio Astronomy Image Archive

On the political level, the period was still dominated by the Cold War. It also witnessed the biggest diplomatic success of the Union: in 1982, mainland China re-entered the Union, after an absence of more than two decades, jointly representing 'China' with a delegation from Taiwan. This was made possible by a breakthrough compromise, involving one country (China), with two adhering organisations with the same name (the Chinese Astronomical Society), headquartered in Nanjing and Taipei.

But for the IAU as an organisation, the period was also one of stagnation. While student protest caused far-reaching reforms at universities and research institutes around the world around 1970, the IAU hardly changed. Of course the IAU was not a university, and it had no students; besides, astronomy was not involved in any of the controversial issues that were the focus of protests: military research, toxic chemicals, or social inequality. There was relatively little pressure to change.

Attendance at the General Assemblies, though still significant, declined in this period. The endless discussions about reforming the commission structure and the General Assemblies remained fruitless. Young astronomers regarded the IAU as a rather closed organisation for elderly scholars, as is testified by several of the interviewees in this book. Reform was clearly needed, but it would only be started after the end of the Cold War, and then slowly.

Science Diplomacy

The period started with something unusual: two General Assemblies in one year. In 1973, the General Assembly was organised in Sydney, the first one in the Southern Hemisphere. But that same year, an 'extraordinary General Assembly' was held in Poland, to mark the 500th birthday of Nicolaus Copernicus. An extraordinary meeting had occurred before, in 1931 in London, but this was merely for technical reasons: to approve new statutes ahead of the actual General Assembly the next year. The extraordinary General Assembly in 1973 was a purely scientific meeting. One can understand that the Australians, who were already worried about the number of people who would be willing to make the long and expensive journey to Sydney, were not amused. Neither was the General Secretary Cornelis de Jager, who had to organise two large conferences instead of one. What had happened?

In this period, a public competition between General Assembly candidates was seen as something to be avoided (just like, say, a public showdown between several candidate Presidents). If there was a serious invitation, other candidates were politely told to reconsider or postpone theirs.[1] In this case, Australia had already been chosen, but Poland, backed by the Soviet Union, refused to scale down its proposal into a Symposium. This put the Executive Committee in a difficult situation, especially because, as somebody wryly commented, the Soviet Union and Poland were not used to public defeats in elections.[2] According to then Assistant General Secretary Georgios Contopoulos (in his interview in this book), the Soviet Union even threatened to withdraw from the Union completely. The compromise to have an extra 'extraordinary' assembly avoided embarrassment on all sides. Besides, many (especially Eastern) European astronomers, who could not afford to travel to Australia could go to Poland—perhaps there was little direct competition after all.

---

[1] See Archives box 6, file 'GA Invitations'.
[2] See correspondence in IAU Archives (old part) I.14b.F.

Fred Whipple at the 1973 IAU GA in Sydney. Photo: Courtesy of the CSIRO Radio Astronomy Image Archive

In the end, the General Assemblies in Sydney and Poland had over 800 participants each. The meeting in Poland consisted of six scientific symposia in Warsaw, Torun and Cracow, but no business meetings. The ordinary General Assembly in Sydney was a full meeting, with a grand opening in the iconic new Opera House. For Cold War reasons, the Sydney organisers expected difficulties with visas for participants from the German Democratic Republic and North Korea, but in the end, participants from Taiwan had the most trouble entering Australia.

Ten years later, the opposite nearly happened: instead of having two General Assemblies, the agreed one in preparation got into trouble. After Grenoble and Montreal, an Eastern European city had been selected: the Bulgarian capital of Sofia (an initial plan considered Varna in the same country). But in 1979, the Bulgarians had to withdraw because of the announced celebration of the 1300th anniversary of their country, which would make the venues unavailable. Besides, since the only astronomical member of the Academy of Sciences, N. Bonneeff, had died, it was hard to keep its financial support.

Several alternatives were discussed, including India and Spain, the US and Greece. In the end, the 1982 meeting went to Patras in Greece, with some additional financial support. The US was standing by in case anything went wrong.[3] The General Assembly was opened in the historical Odeon theatre, triggering many references to the historical origins of astronomy in classical antiquity. It was a success, but with a dark edge, because during the meeting the news arrived that IAU President Vainu Bappu had died in hospital. He was recognised as one of the main founders of modern astronomy in India. The next General Assembly, already planned to take place in Delhi in 1985, became dedicated to him.

Manali Vainu Bappu, here seen at the IAU GA in Sydney in 1973. Photo: AIP Emilio Segrè Visual Archives, John Irwin Slide Collection

The IAU's apolitical profile was crucial for its aim to facilitate relations between astronomers of all nations. But being neutral sometimes also meant that the Union had to refrain from action, even if astronomical interests were at stake. On several occasions in the 1980s, IAU members wanted to speak out against the militarisation of space, but the Executive Committee considered this too 'political' for the Union.[4] As before, political issues, also the

---

[3] See correspondence in IAU Archives box 6, file 'GA Invitations'.
[4] IB 62.

ones involving support for dissidents, were mostly (if not always) referred to ICSU.[5]

The strength of the Union's neutrality was compellingly demonstrated by the most important diplomatic breakthrough in IAU history: the compromise solution to the problem about Chinese membership. The People's Republic had withdrawn in 1959, when Taiwan was admitted under pressure from the American government (see Chap. 2), and all subsequent attempts at compromise had failed. Still, there was broad consensus within the IAU that full Chinese representation was important, if only because astronomical activities in mainland China were significantly more substantial than in Taiwan. The Executive Committee did not want to expel Taiwan, however, as several other Unions had done.

While in 1971, the famous 'ping-pong diplomacy' between the USA and the Peoples' Republic initiated a period of less belligerent geopolitical relations, still during the Chinese Cultural Revolution a solution was beyond reach. But towards the end of the 1970s, personal contacts with Chinese astronomers could be resumed. First IAU General Secretaries Edith Müller and Patrick Wayman, then President Adriaan Blaauw started to discuss IAU membership again.

Incoming AGS Patrick Wayman at the IAU GA in Grenoble. Photo: AIP Emilio Segrè Visual Archives, John Irwin Slide Collection

---

[5] EC minutes 46.

A breakthrough was negotiated by Adriaan Blaauw and Chang Yu-Che, the President of the Chinese Astronomical Society, in the margins of the General Assembly in Montreal in 1979, which was attended by a delegation from mainland China.[6] Any suggestion that there was more than one China, or that Taiwan was an independent country, was still unacceptable, but the 'one China' could possibly be represented by two adherent bodies: one from Nanjing and one from Taipei. After some diplomatic fine-tuning, this—formally temporary—arrangement was officially approved in 1982. Around the same time, a similar arrangement was accepted by the International Union of Biochemistry; Blaauw reported that the two Unions coordinated their efforts.[7] This solution was later also adopted by other scientific unions. This was an obvious case in which a neutral, non-political organisation could bring people together: the core aim of the IAU.

The IAU has been supported by some of the greatest astronomers of their time. Readers will notice that their names appear frequently in this book. Examples are the close personal friends Oort, Blaauw and Strömgren. The photo shows Adriaan Blaauw, IAU President 1976–1979. As President the master-minded China's return to the Union. He later chronicled the first 50 years of the Union and followed the general assemblies with great interest, here in 2006 in Prague. Photo: ESO Archive/H.-H. Heyer

---

[6] This is described in detail by Blaauw (1994) 189–204.

[7] Blaauw (1994) 204.

Cornelis 'Kees' de Jager: 'When the Community Asks …'

President of Commission 36 Theory of Stellar Atmospheres (1961–1964), Assistant General Secretary (1967–1970), General Secretary (1970–1973)

Interview conducted in Alkmaar on 18 April 2017

As the world of astronomy has evolved over time, so has the role of the IAU but one of the recurrent features has been its function as a forum for astronomers to meet, to discuss issues of common interests and to forge friendships across borders, nationalities, political or religious persuasion, etc. Not surprisingly, as they met, scientists harboured opinions, feelings and historical awareness of many divisive contemporary issues. But as their science took the centre stage, it also became a powerful vehicle for international exchange and cooperation. Especially after WWII, the IAU became the institution to provide that stage.

Kees de Jager, photographed in Alkmaar. Photo: C. Madsen

For Kees de Jager it began in 1948 when he attended the first post-war IAU meeting in Zurich. 'For the first time, I met all the "great names" of astronomy. It was fantastic. Otto Struve, Harlow Shapley—I saw him, the great Shapley, I hardly dared to talk to him! For me, that was the most important thing about the IAU.'

Hailing from the Dutch North-Sea island of Texel, Kees de Jager grew up in what at the time was the known as the Dutch East Indies, today's Indonesia. In the early autumn of 1939 he returned to the Netherlands to commence studies as a freshman at Utrecht University, only to find himself and his countrymen embroiled in the Second World War as, in May 1940, the Netherlands was occupied by strong German forces. He studied under Marcel Minnaert, who in May 1942 was arrested by the occupiers. Fearing that a possible deportation to Germany awaited the young de Jager, he took his fate into his own hands. So, like his later friend, Jean-Claude Pecker, de Jager spent a long time during the war 'underground', i.e. living for more than two years in hiding to avoid arrest. His hiding place was a small room at the Utrecht Observatory, quietly made available by Jacob Houtgast, Minnaert's temporary replacement. During that time, de Jager continued his studies, preparing and even partly undertaking examinations in mathematics with Albert A. Nieland and Leon Rosenfeld. 'My first student in hiding,' Nieland said.

(Continued)

A moment of fun: Minnaert and his students rehearsing the reading of Galileo Galilei by Berthold Brecht for a play at Utrechtse Fysisch Lab in the mid-1950s. Seen are (left to right: Jaap Houtgast, lower behind him Hans Hubenet, Mimi Deckers, Jacques Beckers, Corrie Knoppers and—kneeling with a telescope—Kees de Jager. Photo: H. Nieuwenhuijzen

In 1948 he also met Jean-Claude Pecker, who thanks to the new IAU exchange scheme came to the Netherlands to work with Minnaert in Utrecht. 'Since those days we have been very good friends.' De Jager continued his scientific career, obtaining a PhD in 1952 and working at the institute in Utrecht. Then in 1966, he received a telephone call from Pecker, asking him, if he might be willing to become the next Assistant General Secretary of the IAU from 1967 onwards.

'At the time, I was on vacation on the island of Texel. I had been a student of Minnaert, and Minnaert always said "when the Community asks you to help, you should say yes". So I said "yes". It was as simple as that! So I worked with Luboš Perek as his Assistant General Secretary from 1967 and then became General Secretary at the Brighton General Assembly in 1970 when Bengt Strömgren became President of the IAU. In those days, the Secretariat moved with the General Secretary. Just then, we had a new building at the Space Research Institute in Utrecht, so the secretariat was installed there, led by Arnost Jappel.

Before that, however, when Otto Heckmann was President of the IAU, he was approached by Wilhelmina Iwanowska from Torun Observatory in Poland. She said that "1973 will be the 500th anniversary of the Nicolaus Copernicus. I think there should be an IAU General Assembly in Poland on this occasion". But we already had the XVth General Assembly in Sydney, Australia planned for that year. She was a nice lady but also quite strong—when she wanted something, it happened. And she insisted: "Extraordinary General Assemblies can occur in

(Continued)

extraordinary circumstances. And the anniversary of Copernicus is an extraordinary circumstance."

Wilhelminia Iwanowska, the first female IAU Vice President at the Sydney GA in 1973. Iwanowska was also the primary promoter of the second IAU GA held in the same year in Warsaw. Photo: AIP Emilio Segrè Visual Archives, John Irwin Slide Collection

I realised that the Sydney General Assembly already posed a challenge. It was, in fact, the first General Assembly held outside of Europe or North America. For most people, Sydney was far away. There would be few people attending and we should all do our utmost to make it a success. That would be even more difficult with an extraordinary assembly in Poland shortly afterwards. This notwithstanding, Heckmann faced a difficulty. He was, of course, aware of what the Germans had done in Poland during the war and given his own background—like so many German academics he had felt obliged to join the national socialist party in the 1930s—he found it extremely difficult to say "no" to the Polish, so he gave in.'

Iwanowska had argued her case with the anniversary of Copernicus forcefully. But was there also an underlying element of Cold War rivalry?

'I don't think so …, but one never knows. However, the decision caused dissatisfaction in the Community, especially in the USA with prominent American astronomers vowing not to attend further IAU General Assemblies after Sydney. In any event, having taken over as General Secretary at Brighton it thus became my job to prepare two General Assemblies during the same year. In the end, I left much of the preparations of the Warsaw meeting to my successor, George Contopoulos. Of course Warsaw enabled strong participation by astronomers from the East Bloc, and eventually, the waves died down.'

De Jager's original field was Solar spectroscopy and his interest in the Sun has remained unchanged despite 'an excursion' into the field of hypergiant studies. Still today, after having returned to his birthplace, Texel, he is active as a scientist, working on Sun-climate relations at the Royal Netherlands Institute for Sea Research.

Georgios Contopoulos: The Challenge of the Third Integral

President of Commission 33 Structure & Dynamics of the Galactic System (1967–1970), Assistant General Secretary (1970–1973), General Secretary (1973–1976)
  Interview conducted in Athens on 24 October 2017
  A life between stellar dynamics and the defence of science in an unruly world.

George Contopolous in his office at the Research Center for Astronomy and Applied Mathematics of the Academy of Athens. Photo: C. Madsen

Born in Aigion in Greece in 1928, like so many of his contemporaries in the science world, his early years were dominated by dramatic events that could easily have prevented him from pursuing what would become an impressive professional path in science—the occupation by Italian and German troops, the struggle for mere survival in the face of violence and hunger, the subsequent Greek civil war and the long-lasting economic and societal consequences of those years of indescribable hardship.

'I had an interest in astronomy from my early childhood. In high school, I was fortunate that my mother was a mathematician—in fact, the first female mathematics teacher in Greece.

During the Second World War with the German occupation, we had to work in the fields to survive. However, I found the university books of my mother's—before finishing high school—and I sat in the evenings reading about analytical geometry, integral calculus and so on.

To enter university, I had to pass an entrance test. I chose two—one in engineering and one in mathematics. I got the best grades of my group in both tests but decided in favour of mathematics. That was considered a folly since engineering was prestigious. Instead, I attended the Mathematics School of the University of Athens. In 1948—the second—year of my math studies I was offered a position—unpaid, of course—at the Laboratory of Astronomy and from that time on, I devoted myself to astronomy.'

Two years later, Contopoulos was offered a real position at the University of Athens. That also involved carrying out observations at the National Observatory.

(Continued)

'The overall conditions for observations, both in terms of instrumentation, logistics and scope of tasks made me feel I was wasting my time. Instead, I started theoretical work in dynamical astronomy and cosmology.'

In 1953, he obtained his PhD at the University of Athens with a thesis on the motion of stars in globular clusters, published in the Zeitschrift für Astrophysik.

'I attended the IAU General Assembly in Dublin in 1955. Here I met people like Jan Oort and Bertil Lindblad. Lindblad invited me to come to Stockholm, which I did during the next year. He asked me to extend his linear epicycle theory of galactic orbits to higher orders, and I did that to all orders. Then I tried to determine the orbits of stars in galaxies in three dimensions. At the time, people thought that the stars were moving in a chaotic way. My calculations showed differently—that they were orderly, in fact.

At the following IAU General Assembly—in Moscow—I gave a talk on my work. In the audience, Grigori Kuzmin—an Estonian astrophysicist of Russian descent—remarked: "Your calculations indicate that there is another integral of motion in this model!" At the time, I rejected this idea because of a nonexistence theorem of Poincaré. Kuzmin later wrote that in retrospect he had come to agree with me, but meanwhile, I had found a formal integral—the "third integral"—and I sent it to Kuzmin.'

Participants to the IAU GA in Moscow, 1958. In the second row: Geoffrey and Margaret Burbidge, Bart Bok, and Oleg Melnikov. Jan Oort is seated behind Bok and Melnikov. Photo: Oort Archive/J. Katgert

(Continued)

In many ways, the General Assembly in Moscow proved to be a crucial event for Contopoulos. The third integral and the motions of stars in galaxies became an important topic—and stayed with him as a scientific challenge. Also at the meeting, Dirk Brouwer invited the young Contopoulos to come to Yale as a visiting professor. An international career was launched with further stays at Harvard, MIT, Cornell, the University of Chicago, ESO and many other places.

'I went to the following IAU General Assemblies—in Berkeley in 1961 and, in 1964, in Hamburg. At Hamburg, Bart Bok proposed that I should become President of Commission 33 on the Galaxy. It was a great honour. Bok said to me: "I'd like you to take over as President of the Commission, but you've got a misfortune! You're not Dutch". After all the previous Commission Presidents had all been Dutch—Jan Oort, Adriaan Blaauw and Bart Bok. However, at the insistence of Bok, I became the next President of that Commission.

In Commission 33, we organised a number of meetings, but, in particular, I remember one in Thessaloniki on the orbits in the Solar System and in Stellar Systems. It brought together stellar dynamicists and those dealing with celestial mechanics. The two groups had not had dealings with each other before, but here they started having very fruitful interactions.

Otto Heckmann, then IAU President, invited me to come to Hamburg in 1969 to give a talk. On that occasion, he also asked me, if I might be interested in becoming General Secretary of the IAU. It was, of course, a great honour, but I hesitated.'

In fact, Contopoulos was in a precarious situation. Since 1967, Greece had been under a harsh military rule. Contopoulos had spoken out against the dictatorship and intervened on behalf of university students in trouble.

'In a letter to Chandrasekhar, I had spoken my mind. For a while, he didn't answer and it made me nervous. I feared that the letter might have been intercepted.'

As he was to find out, he was indeed under police surveillance and the authorities wanted to silence him. In any event, it would not be easy for him to travel—a requirement for an IAU General Secretary.

'Travelling abroad required a special permission from the police. Even with a permission and even if I was on the plane, I was not sure that I could travel, because I had seen a colleague being removed from the plane at the very last moment.

But Heckmann insisted: "If you have this position, the regime will not dare to harm you!" and so I accepted the job. Thus I began as Assistant General Secretary at the Brighton meeting in 1970.'

A topic that had been discussed for some time was the option of establishing a permanent office of the IAU in Paris. At Brighton, the Executive Committee took the formal decision in this regard.

'Given the problems that I had with travelling, however, that was impossible for me. So I said: "If you insist on that, I will have to resign!" This happened as I had just been elected. Thus the Executive Committee decided to postpone the establishment of the office in Paris until the time that the next General Secretary would take over.

My main task as Assistant General Secretary was preparing the many IAU meetings—about 50 during the three-year period. Before that, also at Brighton, I proposed to establish a European Astronomical Society. The proposal was well

(Continued)

received by the French, the British, the Germans, as well as the Russians. I remember that Viktor Ambartsumian was very much in favour, but when the Soviet astronomers returned back home they were not allowed to participate. In the end, the idea was only realised in 1990. However, with the negative decision by the Soviets, instead, I introduced the IAU regional meetings, the first of which took place in Athens in 1974. Meanwhile, I took over as General Secretary at the Sydney General Assembly in 1973. After that, we went to the extraordinary General Assembly in Warsaw.'

The Warsaw meeting formally marked the 500th anniversary of the birth of Copernicus—in itself an entirely meritorious reason. Yet, the IAU had already chosen Sydney as the venue for its XVth General Assembly, but the push for a General Assembly in Warsaw had come with heavy pressure.

'The Soviet threatened that if we did not agree, they would withdraw from the Union. Thus we decided to have an extraordinary General Assembly in Warsaw. This meeting was quite successful.'

For astronomers, the IAU has historically provided a platform for scientific exchange, but across the great ideological divide of the time, it was not always easy. In the broader context, ICSU had a similar role.

'At an ICSU meeting in 1973 in Leningrad, in which I represented the IAU, we discussed the importance of international collaboration and exchanges of scientists. It was clear, that the Soviets had their own view on this topic. Ambartsumian, who presided over the meeting, declared: "We don't agree with the idea of inviting individual scientists. When we talk about international relations we mean inter-nations" relations, so you have to address yourself to our Academy of Sciences and they will decide, whom to send. Speaking as I could with my own background of coming from a country ruled by the military, I replied: "It is not good if scientists of a country are isolated from the international community. This is the case even for a country as big as the Soviet Union". Ambartsumian and I remained good friends, however. I understood that he had to make this statement even if he may have thought differently himself.'

For Contopoulos, dealing with political interference at times of great insecurity had become an every-day business. In November 1973, the Greek military government was toppled by another group of officers. This was a tense moment for everyone, including Arnost Jappel and Jana Daňková, the two Czechoslovak employees at the IAU secretariat at that time in Thessaloniki. A few years earlier, they had witnessed the Warsaw Pact invasion of their native Czechoslovakia.

'I organised the first regional meeting of the IAU in January 1974. For the organisation, the Greek ministry had given us a lot of money. We published a booklet stating that the meeting was organised by the IAU. Seeing this, the minister became very angry with me. He insisted that the booklet should say "organised by the IAU and the Greek Government". I responded that the meeting was organised by scientific organisations and not by countries. It did not satisfy him, suggesting to me to cancel the meeting. But in the end, he accepted my compromise proposal, saying that the meeting was organised by the IAU with financial support by the Greek Ministry of Education. This notwithstanding, the financial contribution was very important. It allowed quite a participation from Eastern Europe—the Soviet Union, Romania, Bulgaria, Czechoslovakia, etc. Some of them came with no more than one US Dollar a day as subsistence, so the fact that we could provide free accommodation and some meals helped enormously.'

(Continued)

By the middle of 1974, the new military government collapsed, paving the way for the introduction of civilian rule and democratic elections. Once again, Contopoulos made his way to the government offices to secure the necessary funding to secure the continued support for the on-going tasks that came with his IAU function.

'"We're under no obligation to keep the promises of the junta!" the new minister told me. "I do not represent the Junta, but the International Astronomical Union," I replied—to no avail. However, luckily the problem was solved administratively by an alert civil servant on the following day.'

When he took up office at the IAU, Contopoulos had been a professor of astronomy at the University of Thessaloniki since 1967, but in 1975 he moved to a similar position at the University of Athens. Thus the staff of the IAU office moved, too. Meanwhile, the plans for a permanent office were postponed once more since it had become clear that also Contopoulos' successor at the helm of the IAU, Edith Müller, had undertaken efforts to set up a temporary office in Lausanne for the time of her tenure.

One of the issues of the time was the question of convincing China to rejoin the IAU.

'I was trying hard to bring back China. I wrote letters and made various suggestions. The Chinese, on the other hand, also tried to find a solution. But it would still take quite a while. In fact, it took until the time of the General Assembly in Montreal, before the outline of a solution became visible. Nonetheless, the Chinese thanked me for my efforts.'

Contopoulos' term coincided with Leo Goldberg's term as IAU President.

Contopoulos (left) and Leo Goldberg, here seen at the Sydney GA. Photo: Collection G. Contopoulos

'Leo Goldberg wanted to be "an American President"—unmistakeably the "top guy". Officially, the General Secretary is the legal representative of the Union and the relations between the President and the General Secretary were not always easy, but of course, the President was responsible for the overall direction of the Union and this also had its advantages. I remember having to attend a meeting of ICSU in Turkey as the IAU representative. This was just after the Turkish invasion of Cyprus in 1974 and I was not granted a visa. Referring to the

(Continued)

ICSU statutes and the freedom of movement for participants to ICSU meetings, Goldberg intervened demanding that the meeting should be cancelled if I would be denied entry into Turkey. He remained firm in this position, working actively with several academies of science. In the end, the Turkish authorities yielded and I got the visa, just one day before my departure. So together, we found a way of collaborating. In fact, at the closing speech at the Grenoble General Assembly, as we had both reached the end of our terms, Goldberg thanked me "for being very tactful and for educating him in our respective duties".'

Thus ended the term of Contopolous, but as had been the practice, he continued for a three-year period as an advisor to the Executive Committee.

'My involvement ended at the General Assembly in Montreal, but not without a surprise. Upon my arrival, the IAU President told me "we just decided to have our next General Assembly in Greece".'

The plan had been to hold the General Assembly in Bulgaria, only to see a last-minute cancellation by the Bulgarians, leaving open the question of where to hold the next General Assembly.

'Returning to Greece, we asked the Ministry for support, but our request was turned down. Luckily, thanks to Constantine Goudas, the authorities in Patras stepped in and thus, we could issue the formal invitation to the IAU and hold the meeting there in 1982.'

George Contopoulos served as Director of the Astronomical Institute of the National Observatory of Greece between 1975 and 1987, followed by a term as Director General of the observatory until 1993. In 1995 he retired as Professor at the University of Athens, becoming an honorary professor. In 1997 he was elected a full member of the Academy of Athens and in 2012 served as its president. To this very day, he turns up at the office to do his research and supervise PhD students. He never stopped his research on order and chaos in astronomy with applications to various fields of physics. The third integral—in astronomy and the worldly affairs of defending the integrity of science—has never left him.

Richard M West, Scientist and Polyglot Diplomat with a Vision

Assistant General Secretary (1979–1982), General Secretary (1982–1985)
    President of Commission 6 Astronomical Telegrams (1994–2000)
    President of Commission 20 Positions & Motions of Minor Planets, Comets & Satellites (1988–1991)
    Interview conducted in Garching on 30 April 2018
    '*I see the job as IAU General Secretary as being a link in a long chain of individuals. It was a pleasure and an honour to serve in that function. People after me will be the judges if I did things well. Indeed, if I have done so, it makes me happy. If not, at least I did what I could.*'

(Continued)

Richard West at the ESO Headquarters in Garching, Germany. Photo: C. Madsen

Richard West was born in Copenhagen in 1941. He studied astronomy and astrophysics at the University of Copenhagen 1959 and obtained his degree in 1964.

'My father was very good at mathematics and although he chose a different professional career, as a student he had worked for a while at the Copenhagen Observatory. As a child, I took an interest in engineering, but around 14, I turned towards astronomy. A teacher kindly lent me the key to the school observatory— it wasn't much used—and thus I could observe all by myself in the evenings.

When I commenced my university studies, I was allowed to live at the Copenhagen University Observatory site at Brorfelde and there I enjoyed a thorough training in observational astronomy.

My first encounter with the IAU was around 1962/1963 when as a student I sometimes helped with publishing the IAU telegrams—the Telegram Bureau was located in Copenhagen at that time.

In 1963, I attended the European Astronomers Conference at the Nijenrode Castle in the Netherlands. It opened my eyes as regards the need for and value of international encounters in astronomy. During the following year, I experienced my first IAU General Assembly in Hamburg. It was exciting to meet so many famous astronomers, and I especially remember Otto Heckmann, who made a point of spending quite some time talking to us young astronomers.

In those days, a distinguished name in Danish astronomy was Strömgren, referring to both Elis, the father, and Bengt, his son who would later return to Copenhagen after a directorship at Yerkes Observatory. And one of the scientific hot topics was the classification of stars by means of spectra and various multi-band photometric systems, including that invented by Bengt Strömgren. Expertise in this area could be found in the USA, but also in the Soviet Union. During the Hamburg Assembly, an agreement was concluded between "my" professor, Anders Reiz, and Evgenij Kharadze, Director of the Abastumani Astrophysical Observatory in Georgia (then in the USSR), enabling Denmark to send a student there to learn the art of stellar classification by means of objective prism spectroscopy, using the 70-cm Maksutov telescope. For various reasons, the choice fell on me. Apart from my grounding in observational techniques, one may have been that during the preceding three years, I had taken lessons in the Russian language. In any event, I truly enjoyed my stay there—made new friends, obtained an insight into an exciting, but very different culture. I returned

(Continued)

to Denmark with my newly acquired professional skills and, after a follow-up visit to Abastumani two years later, with my wonderful wife! And unbeknown to me at the time—I would later find myself working with Kharadze on IAU Affairs—with him as IAU Vice President, me as Assistant General Secretary.'

In 1970, West joined the young ESO organisation, at the time located in Hamburg, as an aide to the Director General, Adriaan Blaauw.

'Then, in December 1978, Blaauw, who had become IAU President, called me by telephone, asking me, if I would be willing to serve as Assistant IAU General Secretary. It caught me by surprise, not only because I was busy with the ESO Survey of the Southern Sky—which Blaauw himself had initiated—but also because I was only 37 years of age. ESO gave its full support, and so, I could accept the challenge.'

Blaauw's choice was well made. Despite his relative youth, West was a respected scientist, possessed a fair amount of managerial experience, was versed in international matters, spoke 8 languages and understood the arts of diplomacy and networking.

In those years, one of the challenges for international science cooperation was the free movement of scientists when the great ideological struggle divided so much on the Earth.

'Astronomy was largely considered as a peaceful endeavour—devoid of politics and commerce—and for that reason, astronomers were generally respected not only in society at large but also in high levels of state. The privileged position sometimes enabled them to achieve a degree of freedom or grasp opportunities that other groups might have been denied.

For example, I remember a meeting of the IAU Executive Committee in 1981. It took place at the observatory in Abastumani in the Soviet Union. One of the members was IAU Vice President Michael Feast from South Africa, and if there ever were two countries far apart, it was South Africa and the USSR. Obtaining a visa for him was no simple task and it was aggravated by the fact that the observatory was located within a 25-km security zone behind the State border. Even so, it worked out alright, thanks to the IAU and the Soviet and Georgian Academies of Sciences.'

But there were limits. The end of the 1970s and the early 1980s were years of high political tension between the two dominant political blocks—"the West" and "the East". Scientists, who in totalitarian regimes did not toe the party line, often suffered. Their peers residing in democratic societies protested on their behalf, sometimes looking towards the IAU as a conduit for such actions. Often, however, the IAU Executive Committee, referred the cases to ICSU, which had both the authority and stipulations to address these questions, but in some cases, the IAU did try to ameliorate an intolerable situation.

One such case was that of Leonid Ozernoy, of the Lebedev Physics Institute in Moscow. As a Soviet dissident, Ozernoy was denied permission to leave the USSR, even including attendance at the IAU General Assembly in Patras.

'When I was Assistant General Secretary, the delicate political work was of course handled by my predecessor, Patrick Wayman, in coordination with Manall Vainu Bappu, the IAU President at the time. But then, in 1982, the baton was passed to me. Despite the challenges, I was fortunate in several ways: I enjoyed excellent relations with "my" president, Robert Hanbury Brown, as well as with Wayman. Also, myself coming from a Nordic country where pragmatism and

(Continued)

diplomacy are deeply ingrained in the culture, this was useful when it came to dealing with the more critical, diplomatic issues. The fact that I spoke Russian and had many friendly connections in the USSR and its satellite states may have helped, too.

In any event, the IAU has no power in the sense of being able to impose its will on external actors. Whatever influence it does wield must be used wisely and quiet diplomacy is often more effective than strong statements in the public sphere. It is not unlikely that the IAU intervention contributed to a successful closure of the "Ozernoy case"—he was eventually allowed to move to the USA—and possibly of others, too.'

The Patras General Assembly happened under a cloud. It had been organised quickly because Bulgaria had withdrawn its offer to host the Assembly on short notice and the conditions for having the meeting in Greece, including the financial aspects, were challenging. Furthermore, during a stay in Munich, the IAU president, Vainu Bappu, had fallen seriously ill shortly before the meeting. Thus began his presidential address that Vice President Michael Feast read out to the General Assembly on 17 August 1982: 'Circumstances, totally unexpected, and to some extent beyond my control prevent me from being with you during this General Assembly.'

'Bappu was the first IAU President coming from what was then commonly described as "the Third World". He had come to Munich to prepare for the General Assembly but also to rest. He had a heart condition and during his stay, it became clear that he was in urgent need of surgery, which he underwent on 9 August. Tragically, although he received the most expert treatment that Germany could offer, the days later, he died in hospital from post-surgery complications.

Born in 1927, Bappu was of course intimately familiar with British culture. He had received his professional training in the USA and yet he always remained faithful to his Indian roots. He combined the very best that these diverse cultures offer. He was a great scientist, a great philosopher and a great human being. As the news of his death reached us, I interrupted my stay and flew back to Munich, but of course, there was little that I could do.

Back in Patras, on one of the last days of the Assembly, we held a memorial session for him, and I particularly remember a very moving speech by Harlan Smith. In the list of speakers, he simply registered as "a friend".

Regarding Patras, two other things are worth mentioning here. First, the fact that the IAU endorsed the establishment of the International Halley Watch, taking account of the forthcoming perihelion passage of the comet in 1986. Speaking now simply as a scientist, I was extremely pleased about that since the study of minor bodies in the Solar System had become my own preferred field of activity. At the same time, organising a global observational campaign like this in many ways embodied all that the IAU stood—and stands—for. The other one was that with the formal ratification by the General Assembly, the China issue had now been successfully closed with China rejoining the Union under the formula that Blaauw had proposed.

We were obviously excited about the renewed Chinese engagement, but we also felt that it was important to secure a reasonably-sized number of individual Chinese IAU members. The Chinese appeared somewhat reluctant, though, possibly because a high number of individual members would have an impact on their financial contribution to the IAU. Fortunately, during a visit to China in

(Continued)

October 1983, we found a solution that was agreeable to both sides. A cap was put on the Chinese contribution for a while, thus helping the integration of our Chinese colleagues in the worldwide astronomical community.'

If the President is at the helm of the IAU, the General Secretary is master of the 'machine room', the IAU office. Since 1979, the physical location of this office had been located at the small guard house of the IAP in Paris, and as General Secretary, West spent one week every three weeks there.

'Clearly, the work of the General Secretary involves a lot of administration. For example, during my term, I dictated about 2500 letters. However, arriving at the office in Paris, it struck me that the ways of the office were fairly old-fashioned. So we acquired a telex-machine—hardly revolutionary then, but the IAU did not have one!—and somewhat later added an up-to-date PC to the IBM "word processor". I also tried to give the IAU Bulletin a more modern lay-out and less dry look.

In terms of programmatic focus, I felt rather strongly about two topics: Involving more young people in the IAU—not necessarily to the liking of everyone—and bringing in more countries from Asia, Africa and Latin America. One of the tools was the new Visiting Lecturers Programme that became possible following the adoption of the 1983 budget by the Executive Committee at Patras. This programme enabled scientists, e.g. when travelling to distant observatories, to make stops en route and give lectures in places that had not yet developed a strong astronomical base. Early visits involved countries such as Paraguay, Peru and Nigeria. In any event, I thought that it was important, within our limited means, to share opportunities and to give talented young people—irrespective of their origin and personal circumstances—a chance to participate in the great common endeavour that astronomy is. In my view, the new countries and the young people constituted the real potential and the future for the IAU. Of course, the Regional Meetings and the International Schools for Young Astronomers were important undertakings in that context, too.'

In 1982, the IAU became involved in a UN Conference on Space, UNISPACE 1982, contributing with a full-day seminar at the UN premises in Vienna about new opportunities for astrophysical research by means of instruments in space. The IAU meeting was chaired by Bengt Strömgren and was organised with the involvement of Alla Massevich and Richard West, who also took care of the expedient publication of the proceedings. In the following years, the IAU would become an increasingly frequent guest at the UN in Vienna, ultimately as a permanent observer in the UN Committee for the Peaceful Uses of Outer Space—offering expert advice in relevant matters.

The 1985 IAU General Assembly took place in New Delhi. It was the first IAU General Assembly in Asia, very much due to the inspired initiative and determination of Vainu Bappu and his colleagues.

'There may have been some concerns about a meeting outside the usual countries, but the truth is that we had excellent support by our Indian colleagues. That included the availability of the prestigious Vigyan Bhavan Conference Centre right in the middle of the government district. Once again, the free movement of scientists was an issue, but the Indian organising committee, the authorities and the IAU did their best. It is, for example noteworthy, that the Delhi General Assembly was the first time after the formal ratification that China

(Continued)

could participate fully in the meeting—and despite the political tensions at the time between China and India over the Himalayas, it worked out fine.

It had been the plan that the General Assembly would be inaugurated by Prime Minister Indira Gandhi. Tragically, she was assassinated in the year before the Assembly. She was succeeded as Prime Minister by her son, Rajiv, who also attended the opening ceremony of our Assembly.'

Rajiv Gandhi shaking hands with Richard and Tamara West, Hanbury Brown, Menon, and other prominent participants at the GA Opening Ceremony in New Delhi. Photo: IAU Archives

IAU is about astronomy and international scientific cooperation, but it has inevitably also been a forum for broader cultural exchange as can be expected when people from all parts of the world come together.

(Continued)

Richard West and IAU President Robert Hanbury Brown at the GA in New Delhi. Photo: IAU Archives

'From the opening, I remember watching Hanbury Brown, who in many ways was a child of the British Empire—an Englishman born in India—and Rajiv Gandhi. Observing the two in a deep and decidedly cordial conversation was a real pleasure.'

After the Delhi meeting, Richard West continued as an advisor for the following three-year term. Though he later served as President of IAU Commissions 6 (1994–2000) and 20 (1988–1991), professional duties at ESO soon absorbed him. At ESO, he increasingly focussed on astronomy outreach, which he dealt with in the same manner as with his IAU assignment—with dedication and professionalism, but also the humility so typical of him.

Although he published papers on a number of topics, as a scientist, his name is closely associated with the discovery of asteroids—or minor planets, as they were called—and comets, not the least Comet West (C/1975 V1), one of the brightest comets of the century. Among the minor planets discovered by him, he named he named No. 2145 'Blaauw', No. 2147 'Kharadze' and No. 2596 'Vainù Bappu'. The announcement of the naming of the latter—appropriately—was made during the Delhi General Assembly.

In 2005, Richard West was elevated to knighthood by H.M. Margrethe II, Queen of Denmark.

Jean-Pierre Swings: Almost a Family Affair

Assistant General Secretary (1982–1985), General Secretary (1985–1988)
   Interview conducted in Brussels on 19 October 2017
   Can one 'grow into' astronomy? It helps if one is born into it … but the issue is then to have a first name recognised!

(Continued)

Jean-Pierre Swings at the Fonds National de la Recherche Scientifique in Brussels. Photo: C. Madsen

'My father, Pol Swings, was an astronomer who worked at several observatories in the U.S., like Yerkes, McDonald, Texas, and also Mt. Wilson—near Pasadena where I was actually born—where he often observed with the 100-inch telescope. I remember as a child crawling into the spectrograph at Mt. Wilson to mount photographic plates in the plateholder. So I've been in astronomy all my life, without initially being much interested in it—it was just the surroundings. In fact, when I began my studies, I chose physical engineering looking towards space technologies and I was aiming to go into optics. But then, there was an opening at the Observatory in Liège and I joined their team as an assistant.'

Unsurprisingly, Swings Jr. enjoyed the practical observational life.

'I've never done theory! I like to observe, to play with instruments—to put a plate in spectrograph and develop it at the end of an observing night, even observing with small telescopes and going out and look at the sky—not the least at La Silla. I loved it.'

After his Master's degree in space engineering and his Ph.D. and D.Sc. in astrophysics, he spent three years participating in post-doctoral fellowships at JILA (Boulder, Colorado) and at the Hale Observatories. But 'space' never left him. In 1973, coming back from the USA, he was invited by an associate of Roger Bonnet to join an infrared mission definition group for a 3-metre space telescope then under consideration at ESA.

'It never flew, for it was in reality a "hand-waving exercise", but it was interesting to become involved in that.'

Gradually, the contours of a professional life with two distinct paths began to emerge: A scientific one in astrophysics moving from Solar Physics studies, over peculiar B[e] stars to quasars and gravitational lenses, and one focused on 'astropolitics', with heavy involvement in numerous committees in ESA, ESO (including chairing the advisory structure of the ESO Very Large Telescope project and the selection of its site), the European Science Foundation as well as, of course, the IAU. Together with Lodewijk Woltjer, he was also one of the four founders of the European Astronomical Society and served in the Space Advisory Group of the European Commission for its 7th Framework Programme in support of scientific research.

If Swings was 'born into astronomy', one could almost say the same about his relationship with the IAU.

(Continued)

Participants to the 9th IAU General Assembly in Dublin. Photo: Courtesy of the Bibliothèque numérique/Observatoire de Paris

'"My first General Assembly" was the one in Dublin in 1955—the IXth IAU General Assembly. I was 12 years old. Then, my father was IAU Vice President. I'm even in the official photo, sitting on the lawn. Staying with my mother, I did not go to any sessions, though. So the first time, I attended a General Assembly in a professional capacity, was in 1967 in Prague, where I presented a short communication on Solar abundances. At that time, my father had become President of the IAU. Also, his predecessor in this function, Victor Ambartsumian, used to visit my father in our home, so the IAU was a natural part of my—and my father's— life for quite some time.

In 1981, during a stay at ESO in Garching, I received a telephone call asking me if I might be willing to become Assistant General Secretary. I immediately went to see Richard West, then in that post and, as it happened, had his daytime job office in the very same building. Richard convinced me that this could be an interesting task.

I became Assistant General Secretary at the XVIIIth General Assembly in Patras. After the General Assembly, I returned to Liège with my work cut out: Lots of letters to write regarding grants for symposia and colloquia, etc. However, the entire country was on strike. My secretary, Denise Fraipont, was obliged to observe the strike, too. And yet, we found a way. I got into the observatory, grabbed the typewriter and thus equipped her somewhere in Liège so that she could go home and type the letters. We then went to Maastricht in the Netherlands, to put Dutch stamps on the envelopes and sent them. I was indeed lucky to be supported by Mme Fraipont during these years, in addition, of course, to the IAU office staff.'

(Continued)

The IAU has always essentially been a shoestring operation, working on an extremely modest budget. Most of the work has been done for free by dedicated people, not unreasonable for scientists who live and breathe for their interest. But it also included the secretaries of the elected IAU officers. These assistants deserve considerable credit for their hard work, often over many years.

'In 1983, Richard West and I travelled to India to try to advance the preparations for the General Assembly in Delhi. This was the first General Assembly to be held in Asia, but for a while, it seemed to us that things did not move fast enough. It was of course under Richard's responsibility—I merely helped him. In Delhi, we met Ganesan Srinivasan from Bangalore and Mambillikalathil Kumar Menon, a fantastic physicist who was also an advisor to Rajiv Gandhi, the coming Prime Minister of India. Both men were very helpful and efficient, and so the situation improved markedly—the funding came and the local organisation likewise. In fact, I have fond memories of the Delhi General Assembly—the setting, shaking hands with Rajiv Gandhi and so on. Still, a few logistics problems remained, in particular regarding some of the accommodation—the dormitories and hostels for the student participants.

I was mindful of that when I took over as General Secretary at the Delhi General Assembly and had to prepare the next Assembly, this time in Baltimore.'

The changing of the guard, Richard West hands over as General Secretary to Jean-Pierre Swings at the end of the IAU General Assembly, New Delhi. Photo: IAU Archives

IAU General Assemblies normally take place during the month of August, but there have been a few exceptions. The 1985 Delhi General Assembly was one.

(Continued)

Because of the monsoon season, it had been postponed until November, ending on 28 November.

'With the next General Assembly taking place in early August 1988, this meant that I effectively had 2½ years to the job—the Proceedings, the Transactions, Reports—Highlights of the Delhi meeting as well as the planning for the coming General Assembly. The General Secretary normally has three years to accomplish all of this. Again, I was fortunate to have the help of Mme Fraipont and I should also like to acknowledge the excellent advice that I got from Richard West.

Add to this the major change in the IAU Secretariat, because of the sudden—indeed unexpected—departure of Brigitte Manning, who had been running the IAU office for years. This happened roughly 1½ years before the Baltimore General Assembly. As replacement, we hired Monique Orine, who came with strong recommendations, but of course was new to the world of the IAU; she nevertheless did a truly efficient job.

Since the time was so short, I could not really contemplate any initiatives in terms of the IAU's commission structure. But one change was that we managed was to get the IAU President-Elect "on-board" so that he or she would be fully prepared for the task awaiting the person. But my focus was clearly on the General Assembly.'

In many ways, the General Assemblies constitute the heart of the IAU.

'Here, people were happy to learn about the exciting discoveries at ESO, CTIO, at Hawai'i and other places. The invited talks and joint discussions etc. were great. So it was simply the place to be to learn new things and indeed to have face-to-face conversations with colleagues from all over the world.'

And yet, the General Secretary, of course, has additional tasks—for example preparing the IAU Bulletins and also representing the IAU towards the outside world—including at a swathe of bodies with which the IAU is affiliated—perhaps first and foremost ICSU, UNESCO and COSPAR.

'"My President" was Jorge Sahade, who was a very interesting, nice and open-minded person. For some reason, however, he was not particularly supportive of our links with COSPAR. Contrary to him and considering the overlap between astronomy and space science, I thought that the link was important as a means of coordination. But it required some innovative solutions to secure continued relations with that body.'

This highlights an important aspect of the running of the IAU: The relationship between the President and the General Secretary—and in general the sharing of views—is crucially important for the success of the Union's activities. This could be seen as a trivial statement, but in cases, it has required great intra-mural diplomacy.

The Patras General Assembly had formally sealed the re-entry of China after a two-decades long absence from the Union, the basic agreement having been worked out at Montreal and in subsequent exchanges.

'In the aftermath, I had very friendly discussions with China. In fact, the Chinese were keen to increase their involvement in the IAU, including being willing to raise their financial contribution. It was not quite simple, though, because of the delicate balance with the Taiwanese, China Taipei in our newly invented nomenclature. Still, China Nanjing had many more astronomers and so, the Executive Committee could—happily—give its consent in this regard. We also had the first IAU symposia in China—in Beijing and Nanjing, respectively—in

(Continued)

1986, one on observational cosmology and one on the origin and evolution of neutron stars. On the occasion of the former—but also in relation to the discussions about the financial contribution—I visited China. I was received in the Great Hall of the People on Tiananmen Square with many dignitaries in attendance. It was quite moving for me.'

Swings' term as General Secretary ended at the Baltimore General Assembly. This meeting had been supposed to celebrate the Hubble Space Telescope, but in the wake of the Challenger disaster, the launch had been postponed.

'The meeting went very well, I believe. Also, the seemingly trivial accommodation issues that we had encountered at the previous Assembly were avoided. The conference centre was well suited for the purpose. But nothing is perfect: The air condition system at the conference centre was a disaster. In fact, the cooling was so strong that my poor secretary could hardly type on the typewriter, her hands being completely frozen. And an embarrassing moment for me occurred during the opening session. It is a custom that at the opening, the General Secretary presents a list of IAU members deceased during the preceding triennium. When one of the names was called, the person stood up and declared that he was still alive. Good for him, of course, but quite embarrassing for me!'

Following established practice, Swings continued as an advisor to the Executive Committee until 1991.

Today Swings is a Honorary Professor at the University of Liège. He has retreated somewhat from astronomical research, spending time also on some of his other areas of interest, such as history, history of music… and geopolitics.

'Thinking back at my involvement with the IAU, and thinking of the question I put to Richard West as I contemplated the opportunity of becoming an IAU officer, the answer was indeed, yes, it was definitely very interesting and challenging!'

Derek McNally: Managing a General Assembly on Fire

President of Commission 46 Astronomy Education & Development (1973–1976), Assistant General Secretary (1985–1988), General Secretary (1988–1991)
Interview conducted in London on 5 July 2017

Derek McNally at the Royal Astronomical Society in London. Photo: C. Madsen

(Continued)

Despite the drama of the XXIst General Assembly, Derek McNally describes his tenure as General Secretary in more modest terms: 'As IAU General Secretary, I was more or less the last of "the oldies"' And yet, with McNally began the long and arduous process of modernising the structure of the IAU. It would bear fruits several years later, but it is also a process that has never ended.

'I was born in Belfast in 1934. My mother was a primary school teacher, my father a grocery manager—but also a practical mathematician. I came to astronomy by chance. While working on my Master's Degree at the Queen's University Belfast in the mid-1950s, I was, in fact, heading towards nuclear physics, but then, I was taken by Fred Hoyle's book "Frontiers of Astronomy". I thought to myself "that sounds better than heat recovery". I had developed a strong interest in spectroscopy, and as a PhD student, I joined Bill McCrea (later Sir William McCrea) at Royal Holloway College. At RHC I was elevated to be an assistant lecturer in mathematics for my third and final year of my PhD endeavours. I also met a very pleasant female chemistry postgraduate student at RHC. She became my wife! I was then fortunate to be offered a further assistant lectureship in astronomy at University College London Observatory and stayed there for the next 39 years, apart from a fellowship for a year (1963/1964) to work with Subrahmanyan Chandrasekhar at the Yerkes Observatory. My initial area of astronomy was positional astronomy and for many years I lectured on Positional Astronomy to 1st and 2nd-year students before getting down to the astrophysics! By this time my interests had moved to star formation as the early digital computers could quickly handle such computations.

Later, I became involved in the combat against light pollution and other interference with astronomical observation.

Edith Müller was really the person who got me into the IAU "system". Also, Patrick Wayman—the Director of Dunsink Observatory—and Jean-Claude Pecker gave me great support.

My predecessor as IAU General Secretary was Jean-Pierre Swings. Basically, my time as Assistant General Secretary meant "learning the trade". It was less frenetic than today. The volume of publications was not so high, the number of symposia was rather less. At that time, Jorge Sahade was IAU President. He was a very pleasant person to work for and I got along with him very well.'

McNally's term as General Secretary began at the General Assembly in Baltimore in 1988, now serving under the presidency of Yoshihide Kozai.

'He was a charming person, very helpful and gave careful thought to IAU matters. I was very fortunate to have him as President. My first task was one of local housekeeping—organising the move of the IAU Office from the old gate house at the Paris Observatory into the 3rd floor of the Institut d'Astrophysique. Here we found pleasant surroundings and quickly became "part of the house". Also, there were personnel changes at the office: Brigitte Manning, who had served the IAU for quite some time, was just leaving and my predecessor had hired Monique Orine as the new Executive Assistant. Monique appeared to be a competent and efficient secretary and we got along very well. She also kept me well informed of French conditions of employment.

Finances will always be a concern of a General Secretary, and obviously, it was for me too. Adhering countries did not always pay their dues in time while the requests for grants and other expenses, of course, kept coming in. In order to

(Continued)

improve the situation, but in particular, to enable an expansion of our activities, I made a proposal for an IAU Trust Fund. Alas, the Fund did not materialise.'

During its existence, the IAU has seen many adjustments to its Commission structure. Thus Commissions have been dissolved or merged, while others have been created according to the overall evolution of the science.

'Changing Commissions have always imposed a challenge. Often there was significant resistance and the Executive Committee found itself in a difficult position having to "square the circle"—seeing the need for change yet also respecting the wishes of the Commission members. One way forward was "joint meetings" of several Commissions. Another idea, that gradually emerged, was to situate the Commissions within a Division structure. Not surprisingly, it gave rise to considerable discussions.

In any event, I made the suggestion that there should be six Divisions and that an IAU Vice-President should be in the charge of each of them. The Divisions would be introduced three years later, i.e. after my term, but the idea of putting the Vice-Presidents in charge did not fly.'

It has been a constant feature of the IAU that while its charge was to deal with matters of international science cooperation, it had to do so in the broader—often rough—context of international relations. The Cold War period offers a prime case in point. Within its purview, the IAU may have been able to ameliorate the situation, but not always.

'We had planned two colloquia—Nos. 118 and 119—to be held in Nanjing in 1989. Given the return of China Nanjing to the IAU, we had looked forward to having them there. But then came the Tiananmen Square upheaval. In its wake, most international participants withdrew their participation, and thus we felt that we could no longer accord these meetings with the status of an IAU Colloquium.

One of the challenges during my term as General Secretary was dealing with the gender issue within the IAU. There were a lot of quite formidable female astronomers around at the time, albeit less than today. And at the level of the IAU, the invisible barrier had somehow been broken by Edith Müller, who served as General Secretary between 1976 and 1979. I got on very well with Edith and, of course, my successor as General Secretary was Jacqueline Bergeron, so things were improving.

But still, a sort of unspoken resistance clearly remained back then. I did have an encounter with two women astronomers who wanted me to provide a room for a meeting during the General Assembly to advance the number and status of female astronomers. But I deemed that would come under the heading "political" and consequently not a permitted topic, and refused to hold their meeting during the business hours of the IAU General Assembly. However, I did offer to find a room in the evening—an offer accepted rather grudgingly. I am happy, though, to note that a similar evening meeting on Women in Astronomy has been held at every GA since and has proved very successful! It is very satisfying that the number of women astronomers has increased—many such being at the forefront of the astronomy of today.

My major task was the planning of the 1991 XXIst General Assembly in Buenos Aires. The General Assemblies are traditionally held during the month of August, but in this case, the dates of the Assembly were brought forward to enable participants to observe the 11 July total Solar eclipse in Mexico. This was not only

(Continued)

the first General Assembly in South America but also a major event that came not too long after the hostilities between the UK and Argentina had ceased.

Tensions still existed. I remember driving in a taxi with an Argentinian colleague down the Avenida 9 de Julio, the impressive main thoroughfare of Buenos Aires. The driver was jumping up and down in his seat, shouting at the radio, which was turned on. "The Iron Lady", Margaret Thatcher, had resigned as Prime Minister, I then learned from this excited taxi driver.

I did not expect to be well received by the Argentine authorities, but in fact, I was. I had to have a meeting with President Menem himself to discuss the layout of the Opening Ceremony. He did not speak English but had a translator by the name of Mrs. Brown. She was a descendant of the British who had come to Argentina almost a century before to construct the railways. Her English was impeccable and we got on extraordinarily well. She tipped me off that the President favoured astrology and that was likely to give offence to the IAU, but she also said "leave it to me". I did and it was elegantly done. At the Opening Ceremony, Mrs. Brown simply said "astrophysics" in her English translation of the President's speech. She was a true master of the situation. As the government VIPs left the ceremony, she gave me a smile and a wink.'

An icon of astronomy outreach: Patrick Moore not only acted as editor of the GA paper at Buenos Aires, but also gave numerous interviews to Argentine media. Photo IAU Archives

The General Assembly took place in the San Martin Conference Centre which offered plenty of space for the participants, including adequate meeting rooms, a large exhibition area as well as offices for the IAU staff and officers.

(Continued)

The GA on fire, IAU GA, Buenos Aires, 1991. Photo: C. Madsen

'Then, just before the General Assembly was coming to an end, a fire broke out at the conference centre. Fortunately, a lot of the important material was rescued, but we couldn't use the building. While the fire obviously meant an interruption in the ten-day long meeting, it was not "any kind of day". In fact, it happened just as the first results of the Hubble Space Telescope were to be presented to the astronomical community. Also, the final session of the General Assembly with the presentation of the accounts and elections of the new officers

(Continued)

was on the agenda. Not to be rude to the Argentinians, planning ahead may not always be their forte. But dealing with a disaster—give me Argentinians any time. Our Argentine hosts moved Heaven and Earth to solve the problem. Indeed, pulling out all stops, Hugo Levato, Roberto Mendez and their colleagues from the Local Organising Committee handled the situation in a most admirable way. They quickly located alternative meeting rooms, including cafés closed for the winter for small meetings and a theatre for the main meetings. And so, despite the unusual circumstances, things went well and I was fortunate that I was plied with so many soothing cups of coffee by my wife.

After my term as General Secretary, I worked quite a bit with both UNESCO and the UN on behalf of the IAU. Every time I went to UNESCO, I made sure to bring along the most recent colour pictures by the Hubble Space Telescope. At that time, you were not seeing that kind of colour pictures and the UNESCO people were deeply impressed. It greatly helped our access to some key people within the UN system. The subject was, of course, the protection of the skies for the benefit of astronomy, including the issue of a "radio-quiet sky". This was not under the responsibility of UNESCO but was part of the remit of the ITU. Bandwidth meant money, and at the ITU, people were fighting hard to gain the available frequencies, which in turn gave radio astronomy a hard time. For optical astronomy, in 1992, UNESCO organised a conference in Paris on light pollution. UNESCO has followed on up this later on with several international meetings and the topic remains as important as ever, as both radio noise (bandwidth leakage) and light pollution are unfortunately very much with us.'

McNally continued the work on light pollution as chairman of the ICSU Working Group on Adverse Environmental Impacts on Astronomy between 1993 and 1997. Today, he is a Visiting Fellow at the Universities of Hertfordshire and Nottingham studying sky brightness.

Ye Shuhua: Taking Chinese Astronomy from the Ruins of War into the Twenty-First Century

Chair of the Finance Committee (1985–1988), Vice-President (1988–1994)
   Interview conducted in Shanghai on 13 September 2017
   War, civil war, dramatic political developments and the rise of science in China from a rudimentary level to twenty-first century standards—few astronomers can look back at a more extraordinary life journey than Ye Shuhua, the first Chinese Vice President of the IAU (1988–1994), helping to increase international research cooperation.

(Continued)

Ye Shuhua in front of the 65-metre Tian Ma radio telescope in Shanghai, the construction of which marks one of many achievements by her. Photo: C. Madsen

Born in 1927 in Canton, the family moved to Hong Kong when she was 8 years old. In 1942, after Japan occupied Hong Kong, Ye Shuhua moved with her family to the northern part of Guangdong province, where she finished middle school, only returning to Canton after the war.

'I studied at Zhongshan University in Canton. I originally chose mathematics, but during the first year, I was introduced to astronomy. I soon felt that astronomy was more interesting, and thus I majored in astronomy. After my graduation in 1949, and with the civil war still going on, I found myself without a job. I moved back to my family in Hong Kong, where I could find work as a teacher in middle school for two years. In fact, it was the school that I'd myself attended as a pupil.'

In 1951, now with her husband, she came to Shanghai and joined the Zi-Ka-Wei Observatory there. The origin of the observatory goes back to two observatories Zi-Ka-Wei and Zo-San, founded by the French Catholic mission in 1872 and 1900, respectively. In December 1950, the two observatories were taken over by the Chinese government as separate units under the Purple Mountain Observatory. Back in 1926 and 1933, for the First and Second World Longitude Campaign, Alger, San Diego (USA) and Zi-Ka-Wei were chosen to form the Fundamental Triangle to connect the observatories joined in the campaign, and also check the precision obtained. At that time, the equipment in Zi-Ka-Wei Observatory was up to date. After that, because of war, the equipment had become outdated with no way to improve it and it was very hard to maintain normal service.

(Continued)

'Indeed, the initial equipment was poor, but step-by-step we managed to improve and develop better methodologies.

In the new China construction was taking place everywhere, and precise maps were urgently needed. The survey teams worked very hard to meet the demand, but then the poor time service was an obstacle. Thus, an order to improve the Time Service was issued by the State Council to the Academy of Sciences and then to Purple Mountain Observatory because at that time, all the observatories were under the Purple Mountain Observatory. In 1955, key persons were sent from Purple Mountain Observatory to Shanghai, and graduate students from universities and middle schools also joined in.

Also in 1955, ICSU announced that the third International Geophysical Year would take place from July 1957 to Dec. 1958, while the third World Longitude campaign would be included. And then, photoelectric devices were given from Pulkovo Observatory, and Zi-Ka-Wei Observatory was equipped with the new Danjon Astrolabe. In 1957, two sets of grand quartz clocks each containing 3 quartz units were set up there, and in late 1957, the authorities from the Surveying Bureau agreed that the time signal emitted was good enough to use, and thus the Academy had to provide our own UT system, to give the correction of the time signals.'

In 1958, Ye worked with a small group to try to find the way for the Chinese Universal Time Standard. The *Bureau International de l'Heure (BIH)*, cooperating with 39 observatories, constituted the UT standard in the World; the Soviet Standard Time system cooperated with 14 observatories, and Zi-Ka-Wei Observatory joined these two systems from 1939 to 1954 respectively. After working for some months, Ye found the way to enable a small system such as the Chinese one to remain stable over the long term. And so, the Chinese UT system began to be published in 1959.

'The work on the Chinese UT system was very "timely"—if I may use that expression. In mid-1958, ICSU had accepted Taiwan as a member, and all scientific societies in China had to withdraw from the related organisations, including the Chinese Astronomical Society. And afterwards, in 1966, because the political conditions changed again, China had to withdraw from the Soviet Standard Time cooperation, too.

Although the Chinese Astronomical Society could not cooperate with the outside world, the BIH in Paris still very kindly sent their publications to Shanghai, and the Chinese UT system was getting on—first with the cooperation between Shanghai and Nanking. Later on, Wuhan, Peking, Xian, and Kunming joined together, everyone did their best so the Chinese UT system could still be kept in high standard.

In fact, in 1962–1964, when we compared our results with those of the BIH, we saw that our results were better than the Soviets! In 1965, our UT system became the UT standard for China.'

For Ye and her colleagues, after much hard work the future began to look bright.

'But then in 1966 came the Cultural Revolution and everything was turned upside down. It was a disaster. At first, I was still allowed to help with the UT Bulletin work, as an assistant. Later, I was kept away from any scientific work. For two or three years, I worked as a cleaner in the observatory kitchen and later, in Zo-San Observatory as a labourer. Luckily, in 1970, the situation changed again.

(Continued)

In 1971, I could join the scientific planning conference organised by the Academy of Sciences to discuss new plans for the Academy. In those days, space projects were important, new techniques were developed, and laser ranging to satellites was begun in Zo-San Observatory. All space projects needed very precise timing. In 1969, the first VLBI experiments were successful and promised very broad use. And thus, I proposed to build a VLBI system, not only for UT, but also very important for astrophysical research in general. Indeed with our Chinese VLBI network, we're reaping the benefits of this today. The VLBI project was finally accepted by the Academy, and a VLBI group began in 1973 at Shanghai Observatory.'

In 1960, the Zi-Ka-Wei and Zo-San Observatories merged as Shanghai Observatory, directly under the Academy of Sciences.

In 1978, Ye was appointed Research Professor at the Shanghai Observatory, becoming its director three years later and serving for 12 years. And in 1980, she was elected to become a member of the Academy of Sciences, later becoming an academician.

'In early 1979, I was given the chance to lead a team going to the United States of America to learn about the US VLBI and SLR experience, visiting the Goddard Space Centre, the JPL and the McDonald Observatory etc. At this time, very few people were allowed to travel to the West, but the political climate had improved—China had begun opening up.

Also in 1979, I took part in the Chinese delegation attending the IAU General Assembly in Montreal. It was of course a great opportunity to meet fellow scientists from other countries and learn about their work. With the IAU President, Adriaan Blaauw, we discussed the way for China to re-join the IAU. Before, the Chinese position had been the One-China policy, meaning that our country could not participate, if Taiwan did. But things had changed and thus it became possible for both sides to be members of the IAU. Still, finding the proper formula that would enable a solution was not easy. In fact, we could not finalise the process before the end of the General Assembly. Only during the following year could this be concluded and subsequently confirmed at the following General Assembly in Patras in 1982. With China now again a member of the Union, in 1985 I was appointed as Chairperson of the IAU Finance Committee. This was very much due to the encouragement of Richard West, who was General Secretary between 1982 and 1985 and continued for another three years as an advisor. He was a great supporter of me not the least when I proposed to increase the number of grants to astronomers in "poor countries", at the time including, of course, China.'

Around the same time, Ye had become involved in domestic Chinese political affairs.

'I'm really not a political person, striving for high offices, but I tried to do what was best for my country. I did not endeavour to promote any personal agenda of mine or serve a specific interest group, but many people had trust in me and called on me to help sort out a number of problems.'

She was elected as a member of the standing committee of the National People's Congress in Beijing, serving for three five-year terms ending in 2003, as well as being Vice Chair of the Shanghai Municipal People's Congress.

'In 1988, I was elected as Vice President of the IAU. I was the first Chinese Vice President and only the second female Vice President of the IAU. It was a great

(Continued)

honour for China and rather big news in our society. To be honest, I don't know how this came about, but in any event, it started with a letter from Yoshihide Kozai, who was President of the IAU, proposing me to become a candidate. It was completely unexpected for me and first, I did not know how to respond to it, even if my initial inclination probably was to decline. However, having received another letter from the IAU, I took up the matter with the (Chinese) Association for Science and Technology during a visit to Beijing. The Association gave its consent and thus, I could accept the nomination.'

Soon thereafter, on 4 June 1989, the dramatic event in Tiananmen Square in Beijing occurred, creating a chill in the East-West relationship. For the IAU it also meant a challenge. It was only 10 years earlier that Adriaan Blaauw's carefully crafted compromise had paved the way for China's return into the Union and less than seven years since the People's Republic had formally re-joined.

'The political situation in the West and in China are quite different, and it was difficult to reach an understanding. At the Executive Meeting in Stockholm in September 1989, this came to the fore. At issue was, among others, the status of two IAU colloquia, originally planned to take place in China in June 1989 but subsequently postponed. The Executive Committee was minded to cancel these meetings and it put me in an extremely uncomfortable situation, given that I was the only Chinese person present. I opposed it believing that science should be shielded from the rough and tumble world of day-to-day politics. Luckily, with time, the tensions ceased, and I could devote my efforts as Vice President to more mundane matters. It was—and remains—important for China, as indeed for every country engaged in astronomy, to take part in the IAU in order to inter-act with the international community of scientists, regionally as well as globally.'

In 1993, Ye stepped down as Observatory Director and in 1994, her term as IAU Vice President came to an end, too. But she continued to serve in important sci-ence policy functions, including a ten-year term as Vice President of the Chinese Association for Science and Technology and serving five years as President of the Shanghai Association for Science and Technology until 2001. Identifying oppor-tunities and directions for Chinese science in general and astronomy in particular was her focus rather than her own research. And yet she did not lose sight of her own field of science, chairing the international 'Asia-Pacific Space Geodynamics' programme between 1996 and 2008 and as an advisor of the VLBI Group for Lunar Exploration.

After the success of the first Lunar Exploration, the key officers of Shanghai City met the main persons from the related organisation, including Ye and the Director of Shanghai Observatory. On this occasion, Ye asked for a new radio telescope to Shanghai Observatory for the next Lunar exploration. This became a 65-metre telescope developed as a cooperative project of Shanghai City and the Academy of Science. The new telescope was finished in 2012 and has worked very well not only for further space exploration, but also as a powerful tool for astronomical research. Furthermore, upon her suggestion, a planetarium—the largest in the world, in fact—is under construction in Shanghai.

(Continued)

Meeting with Chinese Vice-President Xi Jingping: Seen in the picture from the left of IAU President Bob Williams, Catherine Cesarsky, Norio Kaifu, Jocelyn Bell Burnell, and Brian Schmidt. Former IAU Vice President Ye Shuhua, seen in the photo to the right, was part of the Vice-President's entourage. Photo: IAU Archives

Still at 'only' 90 years of age, Madame Ye, as her colleagues respectfully, even lovingly, call her, gets into the office every day. She has been made Honourary Director of the Shanghai Observatory and is a keen supporter of the continued progress of Chinese astronomy, including being a strong proponent of China's participation in the SKA project as well as the plans for a 12-metre optical telescope.

'China has become a rich country and can support more scientific undertakings, including international projects. At the same time, we need to keep developing our own capabilities.'

Identifying opportunities and directions for Chinese science in general, and astronomy in particular, has been her focus, rather than her own research. Having lived through some of the most momentous periods of China's history and having worked earnestly to improve the lot for its people, looking back at her life, Ye's own way of summarising her contributions could not be shorter or more modest: 'I tried my best.'

Yoshihide Kozai: First Japanese President of the IAU

President (1988–1991)

Interview conducted in Tokyo on 27 May 2017

Even a cursory look at the history of the IAU will reveal a strong imprint on the Union by European and American astronomers. But the IAU belongs to the world community of astronomers and it has underlined this by involving respected scientists—such as Yoshihide Kozai, the first Japanese IAU President—from all corners of the planet.

Yoshihide Kozai, at 89 years of age, in the garden of his Tokyo home. Photo: C. Madsen

Yoshide Kozai was born in Tokyo on April 1, 1928. Like so many others of his generation, the important time of childhood and youth was shaped by the hardship of war. Japan was waging wars between 1931 and 1945, followed, of course, by the deprivation at the times of national recovery from the devastation that the war had brought with it. His personal situation did not make matters easy, even if he came from a reasonably well-off family. His uncle, a philosopher, was a well-known political dissident and that had repercussions on the entire family.

'My father, who was an electrical engineer, fell ill in 1935 and was bedridden for 10 years before he died. During the war, I worked in a metal industry for 14 months, 11 hours each day. Still, before the end of the war, I entered high school. My mathematics teacher encouraged me to read books about astronomy and mathematics, including a Japanese book on celestial mechanics. Later, I came across several French books, like the one by Poincaré on the same subject. It became clear to me that I wanted to become a scientist. And in 1947, I started at the University of Tokyo. In 1951 I joined the Tokyo Astronomical Observatory (of the University of Tokyo) as a research assistant.

When I was a student, I was asked to read papers by Jan Woltjer and getting to know his son, Lodewijk, and sharing time in the IAU Executive Committee in the 1990s was, of course, a great pleasure for me.'

Kozai's main subject remained that of celestial mechanics. In October 1958, he moved to Boston to work at the SAO under the Satellite Tracking Programme and returned to Japan in 1962.

'There were no high-speed computers in Japan at the time, so I decided to work on a topic that would not require such facilities. The choice was high inclination, high ellipticity asteroids.'

(Continued)

That year, he proposed what would become known as the Kozai mechanism for the orbital perturbations of an asteroid caused by other bodies in more distant orbits and expanding on Michail Lidov's work on artificial satellites. This space-age finding has attained renewed importance in the context of exo-planet research.

Yoshihide Kozai in front of a blackboard. Photo: Yerkes Observatory, courtesy AIP Emilio Segrè Visual Archives

'After the war, only a few Japanese astronomers could attend the IAU General Assemblies. It was a dream of mine to attend at least one of them during my lifetime, and the opportunity presented itself at the Berkeley meeting in 1961.

Together with 17 other Japanese newcomers, I was accepted as a member of the IAU at that General Assembly, proposed both by Japan and the USA. I joined Commission 7 on Celestial Mechanics and 20 on Minor Planets. I also became involved with Astronomical Telegrams—Brian Marsden being a friend of mine.

The Berkeley meeting was a great experience. I met many famous astronomers. In particular, I remember meeting Ejnar Hertzsprung.

Ejnar Hertzsprung at the IAU Berkeley meeting. Photo: AIP Emilio Segrè Visual Archives, John Irwin Collection

(Continued)

And luckily, the Berkeley meeting was only the beginning of my association with the IAU and my attendance in General Assemblies. So I went to the subsequent General Assemblies in Hamburg and Prague.

For a while, I acted as a Japanese delegate and in 1988, at the General Assembly, I was elected as President of the IAU. The proposal came from my predecessor, Jorge Sahade. I was the first Japanese national who served in this function, a fact that was duly noted in Japan.

During my time as President, we were keen to increase the participation to the General Assembly and for that, it was proposed to strengthen the scientific content, e.g. by introducing symposia at the General Assembly. This was based on suggestions by a dedicated working group, but at the time, the Executive Committee was still hesitant and so the format of the General Assembly in Buenos Aires remained unchanged, albeit with the decision to enable the following General Assembly to test the idea.'

Meanwhile, as Director of the Tokyo Astronomical Observatory, Kozai in 1988 led the transformation from the university institute into the National Astronomical Observatory of Japan in preparation for Japan's ambitious plans for new large-scale observational facilities and firmly establishing Japan as a main actor in contemporary astronomy of the world. At the same time, the question of cooperation between the countries of East Asia in astronomy remained a prominent issue.

'Already in 1978, together with Norio Kaifu and a few other Japanese astronomers, I went to China to give talks about astronomy—the Cultural Revolution had ended, Mao Tse-tung had passed away and the People's Republic cautiously began to open up. I also went to the Academia Sinica in Taiwan encouraging both parties to resolve their differences regarding the IAU and to collaborate.'

Even if for the IAU, the Continental China and Taiwan issue was settled in the early 1980s, much work remained to bring together the other Asian countries involved in astronomical research.

'In 1987, I chaired the 4th Asia-Pacific Regional meeting in Beijing. The regional meetings have undoubtedly contributed to strengthening Asian cooperation in astronomy. For the meeting in 1987, I invited participants from North Korea and I was very happy that they were able to come. It demonstrates the importance of the IAU as a facilitator that can bridge serious societal divides. Not too long afterwards, during my IAU Presidency, we also managed to have first participation by Vietnamese astronomers.

Of course, the fall out from the student protests in Beijing in 1989 created challenges also for the IAU, reminding us of the delicate balancing act between protecting the integrity of science and the safety of individual scientists.

Already in Buenos Aires, as my term as IAU President was coming to an end, I noted a considerable interest in having a General Assembly in Japan. Like Buenos Aires, Japan was pretty far from "the centre of gravity", but we seized the opportunity and proposed two locations for the General Assembly—in Tokyo and in Kyoto and in the end, the choice fell on Kyoto that then hosted the 1997 General Assembly. The Kyoto meeting was important for Japanese astronomy. The attendance of the Imperial Couple at the opening ceremony also contributed to raising the public awareness of astronomy in Japan. But the General Assembly was equally important for the East Asian region as a whole, attracting many young scientists from China, Korea, and other Asian countries.'

(Continued)

Perhaps the most prolific Japanese astronomer in the twentieth century, Kozai was awarded several prizes and titles, including the Asahi Prize, the Imperial Prize and the Japan Academy Prize, the Brouwer Award, a Second Class Order of the Sacred Treasure, the 'Person of Cultural Merit', and in 2010, he was made an Honorary Citizen of Mitaka City. Sadly, Yoshihide Kozai passed away on 5 February 2018.

Looking Further, Seeing Sharper

Some cynics have said that the space age, which began with the launch of the Sputnik in 1957, ended in 1971 with the last lunar landing. Since then, astronauts have not left Earth's orbit anymore. But for astronomy, the age of space observations was only just beginning. The first generation of space instruments had showed how much there was to discover whenever a new part of the spectrum was opened. Those new instruments were now followed by more systematic surveys, with more sensitive instruments. Both NASA and ESA, and increasingly also national space agencies, launched Gamma-ray, X-ray, Ultraviolet and Infrared observatories, often as international projects. For example, the X-ray satellite observatory Uhuru, launched in 1970, permitted much longer integration times than the short rocket flights, and thus found many more X-ray sources of different and exotic types. Uhuru was followed by a series of increasingly powerful X-ray satellites. The much-anticipated optical space telescope was delayed several times, however, most notably due to the Space Shuttle Challenger disaster of 1986; the Hubble Space Telescope was finally launched in 1990. By the end of the century, space observatories had become routine parts of the astronomical toolkit, albeit highly advanced and *very* expensive.

In the 1970s and 1980s, unmanned probes explored the entire solar system, including the spectacular Voyager missions to the outer planets. In 1986, Halley's Comet was approached by several spacecraft. These missions yielded impressive images as well as an enormous amount of new data. Back on Earth, scientists continued to investigate the samples of lunar soil that the Apollo missions had brought back (Pecker mentioned this as a boundary case of astronomy: did it belong in the IAU?).[8]

---

[8] J.C. Pecker to EC, 22 September 1967, IAU Archives (old part) II.14.D.10.

Launch preparations for the Giotto Space Probe before its flight to Comet Halley in 2006. Photo: ESA

Solar physics got a huge boost as a series of specialised satellites observed the Sun in all possible ways. Solar physics developed into a major sub discipline of astronomy. The entire Solar spectrum could be investigated in great detail—not only in high spectral and spatial resolution, but also its changes over short and long time periods.

One notable new research object was helioseismology: the study of 'sunquakes'—later developing into 'starquakes' or asteroseismology. That the Sun is oscillating with a dominant period near five minutes had been discovered already in the early 1960s by Robert Leighton and colleagues, but the first analysis of them in terms of acoustic waves and stellar structure was only published by Roger K. Ulrich, John Leibacher, and Robert Stein a decade later. From the 1970s, these oscillations could reveal detailed information about the physical conditions in the Solar interior, especially as oscillations with different periods were systematically discovered and characterised. (Serious efforts to observe other stars with enough precision to do the same kind of research only became possible from space after the discovery of exoplanets in 1995, which made exoplanets scientific and politically important. Asteroseismology is now a booming research field.)

The Westerbork Synthesis Radio Telescope. Photo: ASTRON

In 1970, the Westerbork Radio Synthesis Telescope was opened in the Netherlands, thanks primarily to Jan Oort, with twelve 25-metre sized dishes. In 1980, the even bigger Very Large Array in New Mexico, USA became operational—now named the Karl G. Jansky Very Large Array, in honour of the discoverer of cosmic radio waves. Each arm of its Y-shaped configuration is 21 km long. Both telescopes were upgraded several times. By combining different telescopes from all around the world by Very Long Baseline Interferometry (VLBI), even higher resolutions became possible.

The Very Large Array (VLA) at Socorro, New Mexico. Photo: NRAO/AUI/NSF

Radio telescopes also suggested a new way to answer the fundamental question whether we, as humans, are alone in the Universe. Listening for radio signals that could reveal extraterrestrial civilisations became a significant activity. The search, also with other methods (and not only for *intelligent* life) was the subject of IAU Commission 51: Search for Extraterrestrial Life, which was founded in 1982.

While the evolution of stars became better understood in the twentieth century, the first stages of their life were hard to investigate. Some active star forming regions were known, but the formation processes were expected to take place deep inside cold, dense molecular clouds, shrouded behind thick dust clouds. They had to be observed in the infrared spectrum, which required new technology.

Like so much new technology, infrared detectors were first developed for military use (night vision and heat-seeking devices to detect enemy missiles and aircraft). They gradually became available, also for astronomical research in the 1970s and 1980s—first in the 1960s for the near-infrared spectral region, later at longer wavelengths, But it took a long time to develop ones that were sensitive and accurate enough, especially for astronomical research, which required looking at other wavelengths than the military were interested in.

A key problem is the need to remove any interfering 'noise' from the background—including the telescope itself. In order to reduce infrared (heat) radiation from the telescope, it has to be cooled as much as possible: observing in the infrared with a ground-based telescope has been likened with measuring the faint glow of a cigarette stump in the middle of a raging bonfire. During the 1960s the beginnings were made, and the first infrared sources were discovered from the ground, both Galactic and extragalactic. But the Earth's atmosphere proved a formidable obstacle at longer wavelengths, so when space technology had matured sufficiently, in 1983 the first helium-cooled infrared astronomical satellite (IRAS) was launched as a joint multinational project between the USA, the Netherlands, and the UK. It paved the way for further infrared space observatories, which have increasingly benefitted from later technical advances in the form of larger fields, higher resolutions, and higher sensitivity. They made it possible to observe the 'cold Universe', including interstellar molecular particles and dense star forming regions.

The International Ultraviolet Explorer (IUE), launched in 1978 and operational until 1996, not only opened the ultraviolet spectrum for extended research. It was also a breakthrough in the way of working. Until then, space observatories had been planned as missions for specific observations, which were determined by the mission scientists. In the case of IUE, anyone could apply for observation time. If approved, the ground control team would assist with the observations. In other words: as far as astronomers were concerned, the IUE functioned as a 'normal' ground-based telescope, which only happened to be in space. This way of working was also adopted by several later space observatories, most notably the Hubble Space Telescope. It put national and international space- and ground-based observatories on a similar basis— the whole astronomical community could in principle apply for time, but competition was strong.

'Within reach' thus did not mean 'within easy reach': many telescopes are heavily oversubscribed, and the selection process could be tough. In other ways too, competition increasingly played a role in scientific daily life. One effect of this was that ways of publishing changed. The rising cost of journals that were crucial to keep up with the field, and the problems that this caused for institutions from poor countries (or countries with weak currencies) became a concern for the IAU.[9] Peer-reviewed journals became the prime way to publish results; observatory proceedings, which had long played that role,

---

[9] Zdeněk Kopal in GA newspaper 1982, 9.

all but disappeared—and with them, the practice of exchanging them freely with other institutions. In 1970, several national European journals merged to form and fund the new, peer-reviewed journal *Astronomy and Astrophysics*, strongly encouraged by (later) IAU President Adriaan Blaauw.

Optical Telescopes: Renewed Promise

Compared to the spectacular developments in radio and space telescopes, traditional optical telescopes might seem a more sedate field. But building large telescopes was expensive; few observatories could afford to build one by themselves, and unlike radio- or space instruments, large-scale government funding was hard to come by. Pooling national or international resources provided an obvious solution. Two major initiatives were the European Southern Observatory (ESO), in which several European countries participated, and AURA, a consortium of American (mostly East coast) universities, funded by the National Science Foundation (NSF). ESO also became an important centre for European and later international cooperation in astronomy.

One of the important lessons from the early twentieth century was that observatory sites with good seeing were needed to stay competitive, especially in extragalactic research. At the same time, the rapidly increasing urbanisation of California—for several decades offering premier sites—meant that truly dark, competitive optical observatory sites were no longer to be found there. New optical observatories had to be sited elsewhere. Both ESO and AURA started looking for sites in the Southern hemisphere—also to explore new regions of the sky (until the 1960s, most major telescopes were in the North). After extensive site testing campaigns, AURA chose a dark site on Kitt Peak near Tucson in Arizona for their northern national observatory, while a site near the Chilean town of La Serena was chosen for the other, the Cerro Tololo Inter-American Observatory (CTIO). After several years of testing sites in South Africa, ESO also decided to go to Chile, selecting a location 150 km further north, on Cerro La Silla.

The 4.2-metre William Herschel Telescope at the Observatorio del Roque de los Muchachos on the island of La Palma in the Canary Islands, Spain. Photo: Isaac Newton Group of Telescopes

Selecting and preparing the sites and building the telescopes took time, but from the 1970s several international 4-metre class telescopes became operational. Apart from the ESO and AURA instruments, they also included the Anglo-Australian telescope in Australia, the British-Dutch Isaac Newton and

William Herschel Telescopes in the Canary Islands, the German-Spanish telescope at Calar Alto in Spain, and the French-Canadian-American telescope on Hawai'i. Most were located in a small number of places with extremely favourable atmospheric conditions. The largest one was the 6-metre 'Large Altazimuth Telescope' in the Caucasus Mountains in the Soviet Union. It received 'first light' in 1975, and remained the largest optical telescope until 1990. However, because of its location and some technical problems, its scientific impact was less than that of some smaller telescopes. The telescopes of traditional observatories in university towns were taken out of service one by one, and even the best locations in coastal California (Lick, Mt Wilson) were increasingly hindered by light pollution.

The net result was a string of new observatories with names that are still ring familiar today: Mauna Kea and La Palma on oceanic islands; Cerro Tololo, La Silla and Las Campanas on the coastal range in Chile; Kitt Peak, Mt Hopkins, Mt Graham, and Calar Alto on the continent. They also host smaller telescopes, specialised solar telescopes, and other instruments, often operated by several international consortia, which share the basic infrastructure. By 1988, 15 different telescopes were operational on La Silla, for example. A large fraction of the activity of astronomers in the 1960s, 1970s and 1980s was devoted to the development of these observatories and their many telescopes.

The 4-metre optical telescopes became the main workhorses of astronomy. They have been equipped with several generations of new instruments for different wavelengths. Many of the discoveries described in this book were made by the 4-metre class telescopes—often in conjunction with observations in other wavelength domains from satellites, radio telescopes, and other instruments.

In the meantime, technically astute astronomers were beginning to dream of the next generations of optical telescopes. So far, all major telescopes had been built on the same basic principles since the 100-inch on Mt Wilson. A crucial limitation was their price: the cost scaled approximately as the cube of the diameter of the primary mirror. Accordingly, construction of an 8-metre telescope would cost roughly eight times as much as a 4-metre, which was considered prohibitive. Moreover, its size, weight and thermal inertia would probably make its images unstable.

The Multiple Mirror Telescope on Mount Hopkins in its original configuration. Photo: MMT Observatory

New technologies were needed, especially for the primary mirror. New materials and designs could make thinner and lighter mirrors possible. Another idea was to make a large mirror out of smaller segments, although this required developing highly advanced alignment and control systems. That these problems could be solved in practice was demonstrated by the innovative Multiple Mirror Telescope (MMT), consisting of six 1.8-metre actively controlled mirrors, which went into operation in 1979. Its success convinced philanthropist William M. Keck to fund the construction of the 10-metre Keck (I) telescope, which started in 1985 in Hawai'i (operational 1990), closely followed by a second one of the same size. In Europe, ESO followed a more conservative approach. It first developed the 3.5-metre New Technology Telescope (NTT) to test the new alt-azimuth mounting, combining active control of a light-weight primary mirror with an innovative enclosure design. Its first-light observations showed the sharpest images ever seen. That provided the confidence for ESO to begin the construction of a system of four 8.2-metre telescopes with thin, actively controlled monolithic 'meniscus' mirrors: the Very Large Telescope (VLT). Several other 8–10 metre telescopes were also being prepared in this period.[10]

The ESO Very Large Telescope on Cerro Paranal. Photo: ESO Archive/H. Heyer

At the same time, apart from larger telescopes, the quality of observations also improved dramatically by the gradual replacement of photographic plates by electronic (solid state) detectors. After a century of photographic plates, astronomers were keen to improve their sensitivity without losing too much of the advantages of their appealing imaging performance, permanence, and relatively low price. Not surprisingly, there was a parallel military interest in developing sensitive light detectors, for example for satellite imagery and night vision. Military versions of image intensifiers or so-called 'image tubes' were the first to be developed; later versions could be modified for astronomical use. They, and later CCD detectors, reached much higher efficiencies than photographic plates. Another convenience of these devices was the option of real-time viewing and identification of faint sources in the ordinary office light of a control room, instead of at the telescope itself. In the course of the 1980s, they became routinely used by astronomers.

---

[10] McCray (2004) and Madsen (2012).

One very important discovery enabled by these new detectors was the demonstration by Vera Rubin and her colleagues, in the early 1970s, of the existence of dark matter throughout the discs of ordinary spiral galaxies. This was revealed by the constant rotation of the galaxies, which could be measured by the Doppler shift of certain spectral lines (the Hα emission line of hydrogen). The existence of matter, the gravitational attraction of which could be measured by the velocities of known objects, but which could not be seen, had been suggested by Zwicky already in the 1930s, and several indications had been provided by other observations. Rubin's observations were a breakthrough however, and 'dark matter' has been part of standard astronomical and cosmological theory ever since.

Growth and Stagnation at the IAU

The IAU and its Commissions were not directly involved in planning and managing the many new instrumentation projects. It did not operate its own instruments because it did not have the necessary resources, nor did it recommend specific projects. Commission 9 'Astronomical Instruments' also only played a very minor role in telescope development. Commission 50 for the Protection of Existing and Potential Observatory Sites is active in the preservation of existing observatory sites, but has not been prominent in the search for potentially new ones.

Selecting and prioritising projects became a major issue, because national funding was limited. In the US, the National Academy of Science organised 'decadal surveys' to plan the major facilities for the next decade. In Europe, ESO and ESA played such roles in their specific fields. Much later, organisations such as ASTRONET, a network of the major European funding agencies, were founded to coordinate funding for large-scale scientific projects, featuring both scientific organisations and funding institutions. Other organisations, such as COSPAR for space research and SPIE for optical and much other instrumentation technology, did not build specific instruments, but also provided no dedicated forum for coordinating joint instrument projects. In the 1990s, the Organisation for Economic Coordination and Development (OECD) started a 'Megascience Forum' (later renamed 'Global Science Forum'), consisting of government representatives, which held specialist meetings attended by scientists actively involved in the topic. Two such meetings addressed astronomy, and expressed the hope that a global scientific forum for planning astronomical facilities could be established. This does not exist yet, although the IAU is trying to define a meaningful role regarding new international research infrastructures. The task remains difficult; several Commissions and working groups have struggled with it.

The IAU never attained a coordinating role for research infrastructures, but it did provide a forum for exchange of plans for future facilities, such as the Square Kilometre Array (SKA) or (here) the ESO Extremely Large Telescope (ELT), at the Beijing GA in 2012. Photo: C. Madsen

The IAU does, however, provide a platform for scientific insights and informal discussions. As General Secretary Patrick Wayman put it: 'Seldom do IAU Commissions actually guide the procedures adopted in cooperative programmes, but a forum exists where the problems and desiderata can be discussed and the results presented, with a wide range of interested persons being able to take part.'[11] The first Memorandum of Understanding for the giant Square Kilometre Array (SKA) radio telescope was signed at the General Assembly in 2000 in Manchester, for example.

The IAU membership kept growing steadily in the 1970s and 1980s, and the number of adhering countries also grew, adding especially Asian, Middle Eastern and Latin American countries, as well as Iceland. In 1988, a new category of 'associated membership' was introduced for countries in which astronomical activities were still in development, such as Malaysia and Morocco.

---

[11] GA newspaper 1982, 2.

Astronomers from smaller astronomical countries also benefited from the new regional meetings that were initiated: in Europe (from 1974, later continued by the European Astronomical Society), Latin America (from 1978) and the Asian-Pacific region (from 1978, sometimes including Africa); more recently they have been joined by the East Asian Meetings on Astronomy (EAMA, from 1990) and regional meetings in the Middle East and Africa (from 2008). The regional meetings featured scientific discussions, but they also provide a platform for discussing regional cooperation. Other than the high-profile symposia, these meetings had unrestricted participation, which also made them more accessible for early career researchers.

From the IAU Regional Meeting in Bandung in 1981: Venkatraman Radhakrishnan and Joseph Silk. Photo: D. Malin

In other respects, diversity increased very slowly. The General Assembly newspaper of 1985 featured a graph of the birth years of the members: it demonstrated a high peak in 1940, with a very steep decline after that.[12] The percentage of women among the members was still below 10% in the 1980s, although it started to grow. In a telling detail, the General Assembly organisation in Sydney in 1973 included a women's committee for 'women guests who would be attending the IAU Assembly with their husbands'. Among their activities: arranging local wildflowers in the coffee lounge(!). But in the same year, the first woman was elected to the Executive Committee: Edith Müller became Assistant General Secretary, and in 1976 General Secretary. Gradually, attention to inclusiveness increased. The General Assembly in Baltimore in 1988 included informally a session on Women Worldwide in Astronomy, and a Working Group on that topic was later established. In 1992, 200 astronomers (female and male) at the Space Telescope Science Institute signed the 'Baltimore Declaration', which called for a scientific culture in which both men and women can thrive. But overall, the IAU took little action.

In parallel, the diversity of languages used at IAU events also declined. English was the dominant (but never 'official') language after 1945, although not the exclusive one: up to 1967, some formal speeches at the General Assembly were delivered in French. From 1988, the General Assembly programme was no longer published in two languages, but since the IAU legally is a French 'Association', formal documents such as its Statutes and By-Laws still have to be maintained in French. Its official logo combines the acronyms in English and French (Union Astronomique Internationale). It was designed by Jean-Claude Pecker, one of the most active (and one of the last) proponents of using French at General Assemblies.

The IAU logo

---

[12] GA newspaper 1985, 9.

Incoming IAU President Bengt Strömgren addressing the attendees at the Brighton GA 1970. (Left to right) Livio Gratton, Jean-Claude Pecker, Pol Swings, Strömgren, Otto Heckmann, Kees de Jager and Vainu Bappu. Photo: J. Andersen

As mentioned, the General Assembly in Brighton was the first to involve professional conference organisers. With more than 2200 participants, it was the biggest General Assembly so far—and it would remain so until Prague, 2006. Only Grenoble 1976 came close. Clearly, travel times (and cost) was still an important factor; most astronomers still lived in Europe. With concerts by world famous musicians such as Sigiswald Kuijken and Anner Bijlsma, the Grenoble organisers demonstrated that they aimed for cultural as well as scientific excellence. This was not enough to maintain the appeal of the General Assemblies, however. Attendance declined, and criticism on the format of the meetings grew louder.

Future President Leo Goldberg criticised the meeting in Brighton in 1970 as an example of how the General Assembly got out of hand. He referred to earlier discussions: 'not too many years ago [...] there seemed to be a fairly general agreement that the presentation of new scientific results during the General Assembly should be kept at a bare minimum'.[13] New results could better be presented at specialised symposia and colloquia (of which several

---

[13] Goldberg, to Strömgren, 29 September 1970, IAU Archives box 6.

were usually organised just before and after the General Assembly in the same country). General Assemblies should then provide broader overviews of the state of knowledge, and ample room for discussion.

Another complaint that was often made was that General Assemblies were too expensive for the host countries, the Union, and the participants. An attempt to organise a 'cheap' General Assembly in 2000 was not a success, and the experiment was never repeated. Increasingly, external organisations were enlisted: not only the host nation, but also organisations such as NASA, ESA and ESO sometimes supported meetings. The General Assembly in Baltimore in 1988 was also sponsored by five corporate donors, which were involved in astronomical instrumentation.[14]

Several committees looked into the problem, and a questionnaire was sent to the members. That did not help, however. Having shorter, more specialised meetings was generally seen as the way to attract more young astronomers, but the response mostly came from older members, who favoured longer meetings with extensive social functions. In the end, no significant changes were made in the 1970s and 1980s. In the General Assembly newspaper in Buenos Aires in 1991, President Yoshihide Kozai wrote 'I fear that the relative weight of the General Assembly has lessened in the IAU itself as well as in other astronomical societies.'[15] General Secretary Derek McNally was less diplomatic in 1990: he hoped to find a way of 'restoring the General Assembly to the premier astronomical meeting'.[16]

Similarly, the endless discussions on the Commission structure made little progress. Proposals to terminate or merge Commissions always met with much resistance. Successful examples are the merger of Commission 26 with 42; 10 with 12; 4 and 7 with 8; 23 with 24; and the terminations of Commission 17 and 43 (see Commission history in the annex). The same period saw the founding of new Commissions 47 Cosmology (1970), 48 High Energy Astrophysics (1970), 49 Interplanetary Plasma & Heliosphere (1973), 50 Protection of Existing & Potential Observatory Sites (1973) and 51 Bio-Astronomy: Search for Extraterrestrial Life (1982). But until very recently, the fundamental structure remained the same.

---

[14] GA newspaper 1988, 2.

[15] GA newspaper 1991, 2.

[16] McNally to M. Miyamoto in letter 9 April 1990, IAU Archives box 6 file 'Format General Assemblies 1990–2006'.

A significant change was the move of the Minor Planet Center from Cincinnati to the Smithsonian Astrophysical Observatory (SAO) in Cambridge, Massachusetts, following the retirement of Paul Herget in 1978. It was taken over by Brian Marsden, who would be the face of minor planet studies for the next three decades. The MPC was closely connected to the Central Bureau for Astronomical Telegrams, which was also located in Cambridge since 1964 (and was also led by Brian Marsden). Both bureaus were involved in the observation and orbit calculation of comets, for example. Both functioned under the auspices of the IAU, which also provided some (modest) funding. Otherwise, the SAO and NASA provided most funding, and subscribers to the telegram service paid an annual fee.

The Minor Planet Center also coordinated the naming of small solar system bodies. Originally, it mostly accepted names proposed by discoverers, which sometimes included controversial ones (Marsden mentioned that political names could be problematic in cases of regime change, as frequently happened, for example in South American countries in this period[17]). The procedures were gradually formalised, however. In 1978, a small committee was put in place, which later became the Working Group on Small Bodies Nomenclature. In 1988 new rules about controversial or political names were formulated. From now on, political or military figures could only be honoured 100 years after their death or main achievement. In practice, most proposed names were not problematic, however. Many scientists, including still active ones, were honoured with a minor planet.

Coordinating standards and definitions had always been a central task of the Union—the most obvious case in which international cooperation was necessary. Much of that work was done by a small group of specialists. With all the new activities and meetings, it became less visible. Now, some of the most fundamental ones left the Union. The Time Bureau and the International Latitude Service, which also had been co-sponsored by the IAU, were terminated in 1988. Their functions were taken over by, respectively, the *Bureau International des Poids et Mesures* (BIPM) and the new *International Earth Rotation and Reference Systems Services* (IERS).[18] Yet, the IAU remained responsible for defining the celestial reference frame and astronomical terminology and nomenclature.

---

[17] Jana Tichá and Brian Marsden, 'A Short History of the Committee on Small Bodies Nomenclature', in IB 104 (2009) 72–73.

[18] McCarthy and Seidelman (2009).

The first permanent IAU Secretariat was located in the 'old guardhouse' on the premises of the Observatoire de Paris. Photo: C. Madsen

The central office of the IAU itself underwent a fundamental change. In 1979, a permanent office was established in a small gatehouse of the grand historic site of Observatoire de Paris. From now on, the General Secretary would travel to the office, rather than the other way around. As General Secretary Patrick Wayman characterised the premises: 'It is not so big that it will encourage empire-building, but its individualistic setting in historic surroundings is a great attraction.'[19] Wayman would occupy it with the new

---

[19] GA newspaper 1979, 11.

secretary Brigitte Manning, who succeeded Arnost Jappel. Establishing a permanent office was a long-standing wish, but it had been postponed several times, for practical or political reasons (see for example the interview with Contopoulos in this volume). In 1987, Manning was succeeded by Monique Léger-Orine, seconded from the French Centre National de la Recherche Scientifique (CNRS). In that year, the office also moved from the gatehouse to a new office in the buildings of the adjacent Institut d'Astrophysique de Paris (IAP).

As astronomical data was increasingly digitised, so was the IAU administration. In 1984, the office acquired an IBM XT computer for membership administration. But the office remained small, and its facilities limited. General Secretary Richard West reported that much work was also done at this home institution (ESO), for example.[20]

Education and Representation

As mentioned, since 1946, the IAU had started several outreach and support programmes, most notably the *International Schools for Young Astronomers*, which became a flagship activity of the Union.

The other educational programmes remained small. A Visiting Professor project, to bring senior astronomers to places where astronomy needed help, proved hard to realise in practice: only two professors participated in the 1970s (in India and Indonesia). Donat Wentzel, who was involved in many of the IAU's teaching activities, revived it as Visiting Lecturers programme, with exchanges in Paraguay and Peru (a planned one in Nigeria was postponed in 1984). An Astronomy Teacher's Course in Kenya in 1972 was a success, but it was not repeated.

Derek McNally and Richard West launched the 'Travelling Telescope' project, for which they obtained Canadian, and later Japanese, funding in 1988. Its aim was to bring a small, but versatile telescope to developing countries on a temporary basis. The instrument travelled to several Latin American countries and to Morocco, but it was quite expensive.[21]

The General Assembly in Delhi in 1985 featured a special session on popularisation of astronomy. This has been taken as the start of the IAU's involvement in outreach, but its activities remained small, except for outreach projects

---

[20] Reminiscences of Past General Secretaries, IB100 (2007) 12.
[21] John Hearnshaw (S349 forthcoming).

surrounding the General Assemblies. While other astronomical organisations such as ESO or the *Royal Astronomical Society* established professional communication offices in the 1980s—soon to be dwarfed by the hugely successful outreach department of the Space Telescope Science Institute—the IAU did not, probably also because its own staff was still minimal.

The IAU did speak out in public when it concerned protecting the interests of astronomy. In 1973, a new Commission (no. 50) was established for the 'Protection of Existing & Potential Observatory Sites'. In the 1980s, the IAU protested against plans for the 100th anniversary of the Eiffel Tower in 1989, which involved a spectacular spacecraft that would be visible from Earth. Similarly, the IAU objected to plans to use space for advertising, or to send the remains of cremated persons into space. Space debris ('junk') also became an increasing problem. Another way in which the IAU tried to protect astronomy was by adopting resolutions in support of threatened astronomical institutions.

The IAU maintained relations with many other organisations, representing the interests of international astronomy. These ranged from astronomical organisations like ESO, space agencies such as NASA, special committees such as COSPAR (space research) and IUCAF (allocation of radio frequencies), to broader organisations such as ICSU and UNESCO. Since the list was long, the number of IAU officials small, and other issues often seemed more urgent, the level of involvement of the IAU representatives in these organisations varied significantly. It was often dependent on the personal interests and networks of the members of the Executive Committee.

ICSU was a special case, since it was the umbrella organisation, to which the IAU belonged. It could act as the international voice of science, for example in cases when international communication was under political threat, or to protest against discrimination. ICSU could also provide occasional funding, for example for the *International Schools for Young Astronomers* (the actual funds came from UNESCO). The organisation also stimulated specific research programmes, especially in environmental sciences.[22] Even so, the financial relations, and the voting power of the member Unions as compared to the member nations, were a returning topic of discussions between the IAU and ICSU.[23]

In a few cases, the division of labour with sister organisations could be difficult. A rather direct letter from an individual IAU member, who protested

---

[22] Greenaway (1996).

[23] Documents on the relation with ICSU can be found in IAU Archive boxes 30A, 30B, 30C.

against a meeting of the International Association of Geodesy on lunar theory, which he felt was IAU territory, caused some consternation—especially because he invoked the authority of the IAU, which is normally the responsibility of the General Secretary.[24] The division of responsibility between the IAU and COSPAR has also caused occasional discussions.

Each organisation had its own focus, of course. Where IAU meetings will primarily focus on the specific scientific topic under discussion, papers at COSPAR meetings more on the technical aspects of the observation, although the speaker will normally know the science as well. One astronomer observed with some surprise that at a COSPAR meeting, X-rays from a stellar surface and from a cooling galaxy cluster were discussed in the same session. At an IAU meeting, the two contributions would certainly be scheduled in different sessions at least. But when discussing the next X-ray mission at COSPAR, everybody had to share the same data and satellite. The two meetings clearly had different purposes.

A separate, but rather analogous relationship exists between the IAU and SPIE (founded in 1955 as the *Society of Photographic Instrumentation Engineers*, from 1977 known as the *International Society for Optical Engineering*). Detailed technical discussions of on-going and planned projects are generally now held at SPIE meetings, which are large conferences mainly for industry and its customers, but including astronomers engaged in ground- and space-based projects (telescopes, instruments, and detectors for all wavelengths). This division of tasks has actually turned to be quite practical, since SPIE meetings and their combined scientific, technical, and commercial audiences are huge. IAU General Assemblies could never realistically accommodate the technical meetings also, so its focus on the scientific use of existing and future instruments is actually quite appropriate. One of the differences between the pre- and postwar years is that one size (organisation) cannot be made or re-structured to fit all purposes.

The 'Unity of Astronomy'

In 1967, Pecker mentioned 'maintaining the unity of astronomy' as one function of the IAU.[25] As observing technologies proliferated, developments in, say, radio or X-ray astronomy went so fast that it was hard enough to keep up

---

[24] IAU Archives box 14, file 'Commissions (general)': letter from J.D. Mulholland, 15 May 1979.

[25] Pecker to EC, 22 September 1967, IAU Archives (old part) II.14.D.10.

with one subfield. An article in the General Assembly journal of 1979 (where the first results of the International Ultraviolet Explorer and the Einstein X-ray observatory were presented, among many other things) even spoke of an 'identity crisis' in astronomy, as the many new types of celestial objects, and the number of observed sources, stretched nomenclature systems to the limit.[26]

But two decades later, maintaining the 'unity of astronomy' was not an urgent problem anymore. The working methods in various spectral ranges were less different than before (as optical instruments also started using electronic detectors), and there was more interaction between sub disciplines. Combining data from different kinds of instruments had become common. Combining data required coordination of data formats, however—a clear case of 'where international cooperation is necessary or useful'. The IAU contributed by endorsing the Flexible Image Transport System (FITS) data standard in a resolution in 1982; from 1988 the standard was maintained and updated by an IAU working group.[27] These developments culminated in the 'International Virtual Observatory'.

Perhaps the most spectacular example of multi-wavelength astronomy in this period was the supernova that exploded in the Large Magellanic Cloud on February 24, 1987 (SN1987)—the brightest supernova seen from Earth since 1604, visually reaching third magnitude. It was a major theme at the General Assembly in Baltimore in 1988, which had originally been planned to coincide with the first results of the Hubble Space Telescope, before its launch was postponed. Since it exploded, SN1987 has been studied at all wavelengths with every conceivable telescope on the ground and in space in the intervening thirty years (notably by the Hubble Space Telescope since 1990), and a vast amount of detailed information is now available.

Given its brightness, the core-collapse supernova progenitor star was quickly identified. The first surprise was that it was a *blue*, not a red supergiant star as initially expected, so the supernova has been classified as 'peculiar'. Another surprise was that, despite careful searches, no remaining pulsar or any other kind of neutron star has ever been found. However, some 25 neutrinos were detected a couple of hours *before* visible light reached the Earth, probably signalling the onset of the actual core collapse and explosion. This makes SN1987 a prime example of multi-wavelength and even multi-messenger astronomy.

---

[26] GA newspaper 1979, 23.
[27] McCray (2014).

Jaqueline Bergeron: The Quest for Change

Assistant General Secretary (1988–1991), General Secretary (1991–1994)
  Interview conducted in Paris on 2 March 2017
  Change can be a blessing, but not everyone may immediately appreciate it. As IAU General Secretary (1991–1994), Jacqueline Bergeron injected much fresh air, driven by the wish to ensure that the IAU would remain a body seen as relevant to the worldwide scientific community of astronomers. But it turned out to be an uphill struggle.

Jacqueline Bergeron in her office at the Institut d'Astrophysique de Paris (AIP). Photo: C. Madsen

Jacqueline Bergeron was born in France in 1942. In 1966 she received her Diploma in Physics.

'Studying at the Ecole Supérieure de Physique et Chimie in Paris, and still considering which path within physics to take, one of my teachers, Pierre-Gilles de Gennes, suggested that I "go and talk to James Lequeux". After a conversation with him, I was convinced that astronomy was right for me. Many people were doing stellar astronomy, but I wanted to do extragalactic astronomy. With a grant from the CEA in Saclay and Evry Schatzman as my supervisor, I went on to study at the University Paris VII (Diderot) for my "Doctorat de 3e cycle", which I completed in 1968 with a thesis on the X-ray background emission. Subsequently, I received my "Doctorat és Sciences" in 1972.'

Between 1969 and 1980, Bergeron was on the staff of the CNRS, but she also spent time at institutes abroad, in the UK (Cambridge), three years in the USA (Cornell University, Caltech), and finally at ESO in Geneva, before returning to a position at the Institut d'Astrophysique de Paris.

(Continued)

'It was Jean-Pierre Swings who came to ask me if I would be interested in becoming Assistant General Secretary. I was very surprised, but he explained that he thought the IAU needed younger people and indeed also more women. To be honest, the IAU did not mean much to me at the time, except that it appeared to me as an "oldish" institution, but Jean-Pierre was very convincing and so, I accepted.

On accepting the nomination for Assistant General Secretary and, of course, subsequent General Secretary, I felt that we had to do something, if the IAU were to survive. Thus, bringing younger people into the IAU was really important to me. For example, for that reason, I insisted on lowering the fee both for participating in the General Assemblies, but also for the IAU symposia taking place outside of the General Assemblies when I became General Secretary. I was also disturbed by the fact that many countries appeared to be hesitant to nominate young scientists as new members.'

Bergeron became Assistant General Secretary in 1988 in Baltimore.

'Jean-Pierre Swings and I had discussed the need for strengthening the weight of the Union in international relations, not the least in ICSU. ICSU could handle many international issues—some quite tricky—in a way that no other body could do. I really found ICSU to be of great value. As a matter of fact, if it didn't exist, one would have to create it anew. The other burning issue, as already mentioned, was how to attract young scientists to the Union. At the time, at the General Assemblies—in terms of visibility the cornerstone activity of the Union—we still had a focus on small one-day Commission meetings with narrow topics and, frankly, less focus on really interesting science. Indeed what really struck me—and confirmed in many conversations with colleagues—was the need for a scientific format that was interesting to young people, including having real symposia in connection with the Assemblies. However, initially, my suggestions were not met with much love. Many people resent change, and so it was an uphill struggle. It was, in fact, only after taking over as General Secretary at the 1991 General Assembly in Buenos Aires, that I could start a successful lobbying activity for a change. Luckily, in that, I found strong support from Lodewijk Woltjer, then President-Elect, and so, we could change the format of the following General Assembly—to be held in The Hague in 1994—and include six four-day symposia.'

(Continued)

Renewal and tradition: The General Assembly in The Hague brought rejuvenation to the GA proceedings. But it maintained the tradition of a grand Opening Ceremony. The picture shows the opening session. In the front row, with hat: H.M. Queen Beatrix, flanked by (left) Ed P.J. van den Heuvel, Chairman Netherlands National Committee for Astronomy (NCA), and (right) Hugo van Woerden, Chairman National Organizing Committee IAU-GA1994. At far right Adriaan Blaauw, IAU President 1976–1979. Photo: Collection Hugo van Woerden

A first step had now been taken. But picking up on McNally's earlier failed attempt, Bergeron also pressed for a reform of the Commissions—and, once more, encountered heavy resistance.

'Dealing with commissions was really overdue. For example, we had three Commissions on multiple stars, yet we needed only one. Since it had shown to be so difficult to close commissions even when they had lost their original rationale, one idea of mine was to limit the time of commissions to three years, with a one-off renewal, like for the working groups. Yet, it didn't fly. However, this was not simply a question of cutting down. I was lobbying for a decrease in the number of commissions to take out redundancy but also, in fact, to create room for new ones. For example, we had no commission on active galactic nuclei.

Some people resisted the suppression of committees on the grounds of the perception of prestige in some countries by having a national as IAU Commission President. Among them was the IAU President, Alexander Boyarchuk, who was strongly opposed to changes. And so, despite the help of Lodewijk Woltjer, at the time we did not succeed in this quest. The merits of the national prestige

(Continued)

that might be associated with Commission Presidencies notwithstanding, preserving Status Quo was hardly a recipe for successful future for the IAU. And at least, although we did not succeed to change the format of the Commissions, we took an important first step towards a resolution with the proposal to reordering the commissions under a new system of topical divisions. Eventually, the division structure paved the way for reforming the commissions, though it were to take many more years.'

Jacqueline Bergeron was the second woman, only, to become IAU General Secretary.

'It did play a role for me in accepting the job, but personally, I never encountered a problem in that regard.'

Still, it was time to break new ground. For the Buenos Aires General Assembly, Bergeron's predecessor, Derek McNally, had turned down a proposal for a dedicated meeting on the role of women in astronomy. He had not done so for lack of sympathy, but because he considered it 'too political'. But times had changed and with support by Bergeron, and indeed of the IAU Executive Committee, the first formal meeting of its kind could take place at the 1994 General Assembly in The Hague.

Preparing the following General Assembly in Japan was the job for Bergeron's successor, Immo Appenzeller. But the decision about choosing the location had to be taken by the Executive Committee in 1993, i.e. during her term. The reason was that while the original invitation by Japan was to hold the General Assembly at Makuhari in the Tokyo area, Kyoto had emerged as an option, too.

'Having visited the two sites, there was no question for me, that Kyoto should be selected. As the old cultural centre of Japan and with the availability of the Kyoto International Conference centre, this was by far superior to the Tokyo location, which was rather isolated on a newly reclaimed stretch of land close to the airport and with no cultural attractions. But I had to convince the Executive Committee, which included Yoshihide Kozai, the Director of Tokyo Observatory. After some debate, the Executive Committee followed my recommendation, and I still believe that the Kyoto meeting was one of the most successful and memorable IAU meetings.'

The General Secretary is, of course, also responsible for the running of the IAU office, overseeing the small office staff in Paris.

'I tried to simplify operations as much as possible. For example, to avoid travelling and handing out large amounts of cash at the General Assembly, I managed to have a bank desk on-site to dispense travel grants, based on vouchers that we issued.'

From today's vantage point, how does Bergeron see the evolution of the IAU?

'A continued challenge for the IAU, it seems to me, has been the constant battle with itself to maintain its relevance in a rapidly involving world of astronomy, and the Commission issue has been omnipresent in this struggle, both as a vehicle for its interactions with the scientific community at large and as an essential organisational pillar. In this context, the need to attract young scientists and have them engage with the IAU, as members but also serving as IAU officers, remains vitally important.

At the same time, in this day and age the observational facilities for astronomy—space-borne and ground-based—have increased strongly and in parallel, there has been a proliferation of meetings and conferences. Thus we need to

(Continued)

understand better, how our worldwide community is actually functioning, i.e. the link between the number of astronomers, the facilities at their disposal and how in this increasingly complex environment we communicate with each other. To help understand this might possibly also be a job for the IAU.'

After her term as IAU General Secretary, Bergeron took up a position as Associate Director for Science at ESO, until returning to the IAP in 2001. While having in the meantime participated in numerous international boards, panels, committees, etc. she has retained her affiliation with IAP and CNRS, now as 'Directeur de recherche émérite'.

Immo Appenzeller: Progress, Not Revolutions

Assistant General Secretary (1991–1994), General Secretary (1994–1997)
  Interview conducted in Heidelberg on 13 December 2017
  The first IAU homepage, the introduction of the IAU divisions, a successful IAU General Assembly in Asia—only the second in that Continent—and facing the astronomical fall-out of the collapse of the Soviet Bloc. Those were some of the items on the menu for Immo Appenzeller as he took the helm as IAU General Secretary in 1994. 'It was fun,' he says, 'I wouldn't have missed it.'

Immo Appenzeller in his study. Photo: C. Madsen

'During school, I got hold of a popular book about Einstein and his theories entitled "Einstein und das Universum". It caught my interest. In 1959, I began university studies in Tübingen in nuclear physics which was then very much in fashion. In 1961, I moved to Göttingen to study with Arnold Flammersfeldt, who had been a co-worker of Otto Hahn. However, he could not take more students, and because of my financial situation I needed to move on without delay, so instead I went for a diploma in astrophysics—which at the time nobody wanted to do! My chosen topic was a study of local interstellar magnetic fields by means of optical polarisation measurements.

For this, I had a polarimeter built by Alfred Behr, my supervisor. The weather in Göttingen is not suited for astronomical observations, so I contacted the

(Continued)

Haute-Provence Observatory, which was ready to grant me observing time. But to bring my polarimeter to France, I needed permission by Behr, who at the time was in the USA. Instead, Behr suggested that I'd come to Chicago as a technical assistant to the Yerkes observatory. I could do my observations there, but would also need to do some functional work. Thanks to a Fulbright Scholarship, it all became possible. The functional tasks, in fact, involved a computer programme for reducing polarisation data, which I'd already done in Germany. After a short visit to Germany to finish my PhD, I returned to Yerkes to build a new polarimeter for them and a copy for the Kitt Peak Observatory. I could have stayed, but this was during the height of the Vietnam War, and applying for a green card would have implied the risk of being drafted as a U.S. soldier, so I decided to return to Europe.'

During his stay in Germany, Appenzeller had helped Rudolf Kippenhahn with some programming, and Kippenhahn had expressed an interest in hiring him, should he decide to come back from the USA. And so, in 1967, he was able to join Kippenhahn's theory group at Göttingen.

'This gave me my first experience with observations at ESO, which at the time only operated two telescopes, the 1-metre and 1.5-metre, at La Silla in Chile. I was back to observational astronomy, and this became my destiny—observing and instrument building—for Calar Alto observatory, for ESO, and others. This is what I've focussed on during my career, from 1975 as professor of astronomy at Heidelberg University and director of the Landessternwarte Heidelberg-Königstuhl.

My first encounter with the IAU was the General Assembly in Brighton in 1970. I had attended meetings at the German Astronomical Society (A.G.) and the American Astronomical Society, and I liked these large, multi-field meetings. I'd also become more involved in international astronomy, partly based on my work in Europe and the USA, but in 1972 I became a guest professor at the University of Tokyo and later also worked with the Mexicans.

I had worked in several IAU Commissions, notably on stellar structure and stellar spectroscopy and I'd co-organised a few meetings of the IAU, including Symposium 122 on Circumstellar Matter in Heidelberg.

Then, in 1989, I was asked by Derek McNally, whom I knew from Chicago, if I would be interested in becoming General Secretary of the IAU.

I was actually quite surprised about this request, but felt I had to accept the challenge for two reasons: One was the realisation that over time, Germany had not been strongly represented within the IAU. This had much to do with history, but at the same time, we owed much to the IAU because of its role during the Cold War. For decades, the meetings and events organised by the IAU was the only way that we could keep contact across the great political divide, which not only cut through Europe but very much through my own country. The other reason was perhaps a selfish one—that the office was located in Paris, in my view one of the most interesting places in Europe!

In any event, the IAU appeared to me as an important organisation, which could make the opinion of astronomers heard internationally and it could speak with authority.'

Appenzeller became Assistant General Secretary at the 1991 General Assembly in Buenos Aires.

(Continued)

'Formally, my task was to look after the symposia and colloquia, taking care of the applications, preparing them for decision by the Executive Committee, including clearing up any competition, where several proposers were aiming for similar things. That took most of my time, but the Assistant General Secretary also participates in the meetings of the Executive Committee and so, should he or she wish, they can influence decisions at the IAU. At least I, for one, was never shy about expressing my views.'

In December 1991, the Soviet Union collapsed and around the same time, the Yugoslav Republic descended into a fierce and bloody war and fell apart, too—with first Slovenia, then Croatia ceding the federation. Shortly thereafter Bosnia-Herzegovina, too, broke loose, leaving a rump state of Serbia, at the time still with Montenegro and the autonomous Kosovo region. Serbia became the subject of a comprehensive United Nations boycott that also included scientific cooperation. These events obviously posed considerable challenges also for the IAU.

'First of all, the events led to an immediate and strong weakening of astronomical research in this part of the world, because some institutes and the number of active scientists went down. Also, e.g. Russian people working at the Crimea Observatory suddenly found themselves in another country, Ukraine. From an IAU perspective, Russia was lucky in the sense that it could continue in the IAU. This was much more difficult for some of the newly independent states—including, for example, the Baltic countries—many of whom were very poor. Bringing them back into the IAU took a long time. In any event, it was helpful to have a Russian IAU President on the Executive Committee—Alexander Boyarchuk—who played a positive role in dealing with these challenges.

Croatia was the first country from the former Yugoslavia that joined the IAU. Serbia, that still saw itself as Yugoslavia, was under boycott, and re-admission had to wait until ICSU recognised the country in its new incarnation. Macedonia was also a challenge, because the Greeks opposed the notion of an independent country by the historical name of a Greek province. At the General Assembly in Kyoto, the parties found a solution that followed the internationally agreed pattern and the country was listed as the Former Yugoslav Republic of Macedonia—FYROM.

On some of the more difficult political questions, it was helpful that the IAU could lean on the decisions or practices of ICSU, and in interactions e.g. with the UN, ICSU carried considerable weight because it represented all sciences. ICSU was also quite useful regarding the proposed space advertisements in that came up at the time, and we strengthened our own links with the UN-system by becoming observers to the UN Committee for the Peaceful Uses of Outer Space (UN-COPUOS) in 1996.'

The issue of space advertising which came up in 1993 was the 'Space Billboard Project', a giant illuminated billboard in low orbit, obviously meant to be visible from Earth. The project was abandoned in 1993, partly because of protests from science, partly also for major legal, technical, and financial problems.

The 1990s were marked by the emergence of several large such projects, which raised the question of the role, if any, of the IAU in these developments. For example, optical astronomers engaged in the various new 8–10-metre class telescopes began meeting in an informal '8-metre club' outside the auspices of the IAU.

(Continued)

'We initiated a working group on large facilities in astronomy, initially chaired by Harvey Butcher. It's clear that in regard to the cost-intensive infrastructures needed in today's astronomy, the IAU can only have a limited role, since these projects are decided at a different level. Yet the IAU can serve as a useful forum for the exchange of ideas and, where necessary, help to avoid potentially damaging friction between potentially competing projects and explore coordination options, an important aspect in the current "multi-messenger" observational paradigm.

More generally, the IAU was—and is—always developing and that is indeed one of its strengths. Indeed, Jacqueline Bergeron, the General Secretary in office at the time, correctly felt that the existing commission structure had its deficits with a considerable imbalance between the various commissions. Various suggestions for change were made, but they were not successful and met with quite a bit of opposition. At this stage, Lo Woltjer, who was President-Elect at the time, stepped in, with a proposal modelled on the AAS system, suggesting a division system as an overarching structure for the commissions.

My relations with Lo Woltjer were excellent. I knew him already from Chicago, and later also worked him in the context of ESO. The suggestion of Divisions was presented to the General Assembly in The Hague in 1994 when I took over from Jacqueline, and so it became my task to implement the new structure. It was also clear that the new structure brought with it the need for changes to the Statutes and By-laws, something that I started to work on already as Assistant General Secretary.

At the Kyoto General Assembly, the transition was completed.

IAU President Lodewijk Woltjer opens the IAU GA in 1997 in Kyoto in the presence of Japan's imperial couple. Photo: C. Madsen

(Continued)

I was pleased that we could have the 1997 General Assembly in Japan. General Assemblies have been held in many places in the world, but until Kyoto, they had all been in Europe, the United States or in countries that had strong historical links with Europe—in a sense were "Europeanised". Japan was different. It had adopted many European ways, but at the same time, it retained its original thinking and feeling of things. And so it was particularly interesting to hold a General Assembly in this country and also give credit to the progress of Japanese astronomy, not least in radio astronomy.'

At the time, however, Japan had a terrorist problem due to the cult of Aum Shinrikyo. In March 1995, a sarin gas attack was launched on a subway station.

'The Japanese organisers were keen to have the Emperor open the General Assembly, but there were great concerns about his security. Therefore the plans were kept secret. Some of the organisers had to be informed, but the circle of in-the-knows had to be small. One day, Daiichiro Sugimoto, who chaired the National Organising Committee, showed up here in Heidelberg to tell me about the plan. We agreed that initially, only Lo Woltjer, Monique Orine and I were allowed to know; only later, the group of people were extended to those who were supposed to meet the Emperor face-to-face before the opening ceremony. During the time of preparations, we, therefore, had to use a code word for the Emperor when we communicated with each other. In any event, the Japanese did a marvellous job in organising the meeting.'

IAU GS Immo Appenzeller, IAU President Lodewijk Woltjer and Keiichi Kodaira (NAOJ Director General) at the 'Kagami biraki' (opening of the sake cask) cere-mony. Photo: Collection Immo Appenzeller/H. Zinnecker

At 'Kyoto', Appenzeller passed the IAU baton to Johannes Andersen.

(Continued)

'In terms of "domestic matters", one of the first things that I implemented as General Secretary was the IAU homepage. Because I was already used to writing HTML, I coded everything myself. Establishing the website was important, since not only did it signal our use of modern technology, but also because of the speed with which e.g. the Information Bulletin could be published.'

It should be remembered that the first widely available web-browser had been introduced only in late 1993. Perhaps easy to overlook today, for the IAU to go 'on-line' only a little more than one year later constituted an achievement on its own.

'Returning' to science and science management at a national level after his stint at the IAU, Appenzeller could again focus on his research interests, which had expanded to include young galaxies, quasars and cosmic X-ray sources—and of course astronomical instrumentation.

Appenzeller retired in 2005. In 2001, he was awarded the Lussac/Humboldt Prize, followed by the Karl Schwarzschild Medal of the Astronomische Gesellschaft in 2015.

How would he himself summarise his time as IAU General Secretary?

'It was fun to do, it was interesting. Serving as General Secretary takes a significant amount of time, so I had to reorganise my life during those three years—but then, it was only for three years. In a way, I was lucky, too. Aside from the Yugoslavia issue, there were no huge political challenges to affect the work. Other General Secretaries have been less lucky. So for the IAU, it was not a time of revolutions, but it was a time of progress.'

Johannes Andersen: Boy Scout Whirlwind

President of Commission 30 Radial Velocities (1985–1988), Assistant General Secretary (1994–1997), General Secretary (1997–2000)

Interview conducted in Copenhagen on 9 June 2017

'Ever the boy scout, I overdosed from Day One.' Those were the words with which Johannes Andersen, General Secretary from 1997–2000, described his term in office in the IAU Bulletin No. 100. The self-appraisal contains some truth—dedicated, dynamic, with stamina and always forward-looking as he is. But also someone with strong opinions and ready to argue his case.

(Continued)

Johannes Andersen in his office at the Niels Bohr Institute in Copenhagen in front of his favourite picture, an original painting of Danish astronomer Ole Rømer. Photo: C. Madsen

'I was born on 29 August 1943. On that day, the Danish government resigned, a general strike broke out, the occupying power declared a state of emergency,' he says with a smile 'and I was a boy scout for 15 years'. Thus the tone is set.

'During a visit to my grandmother at 14 years, a local amateur lent me his 135 mm refractor. Jupiter and Saturn were high in the clear sky, and I was sold immediately. The next year, I got Fred Hoyle's book "Frontiers of Astronomy" (in Danish) and worked my way through it—with Bohr's atomic model, nuclear synthesis, "steady state" and everything else—until I fell asleep, totally exhausted. It was tough for a 14-year boy, but I've basically never looked back.'

Indeed, looking back is not in his nature.

Andersen graduated from Copenhagen University in 1969 and became a well-known scientist. He was a frequent observer at ESO's La Silla Observatory, using both ESO and Danish national telescopes.

'During a stay at the ESO Guest House in Santiago, I received a call from Jacqueline Bergeron, then IAU General Secretary, asking if I'd be willing to become IAU General Secretary. Having checked with the Director of the Niels Bohr Institute, I accepted.

As usual, I wanted to change everything, but had to wait. The golden rule is that the Assistant General Secretary is seen, but not heard, for there can be only one captain on the bridge at a time. Then, in 1997, at the General Assembly in Kyoto, I took over.

Back in Europe, I started with the basics. The IAU office at the top floor of the Institut d'Astrophysique cried for refurbishment. The Information Bulletin had remained unchanged for many years and received a face-lift from Lars Lindberg Christensen, who was then at the Tycho Brahe Planetarium in Copenhagen, but later became IAU press officer. We also updated the IAU website. And the Rules for Scientific Meetings got a brand new section on gender balance!'

(Continued)

'Around Christmas 1997, Monique Orine, the Executive Assistant and only full-time employee of the IAU, first fell ill and then suffered a vicious dog attack. Knowing what was later discovered, I've often wondered about the background, and if there might be a link with those incidents. I had no suspicion at the time, but at end of my term, I did wonder about some—in my view—questionable expenses. However, this happened at the run-up to the next General Assembly, when a General Secretary is extremely busy, and I left the issue with my successor.'

In the 1990s, discussions about potentially hazardous Near-Earth Asteroids began to reach the public sphere. Whilst a threat to the Earth could not be dismissed, the discussion was not only about safety issues—the Cold War had ended, but some bewilderment might linger in military circles—but was saddled with speculation that this might ultimately require military means, perhaps including nuclear weapons.

'As General Secretary, I was an observer at the meetings of the Scientific/Technical Subcommittee of the United Nations Committee for Peaceful Uses of Outer Space (UN-COPUOS), where the Near-Earth Object (NEO) topic had popped up. In February 1998, just after the first meeting, we discussed the matter among the IAU Officers. I found it unacceptable that the IAU did not have a policy on Near-Earth Objects (asteroids or comets). Everybody else considered NEOs either a scam or crucially important, but I felt that the IAU could not just turn its back on the issue.'

Observations of asteroids are reported to the IAU Minor Planet Center (MPC) in Cambridge, Mass. and entered into a database there. Then, in March 1998, the British astronomer Brian Marsden and head of the MPC issued an IAU Minor Planet Circular suggesting that Asteroid 1997 XF11, a NEO about 2 km in size, would come close to the Earth in 2028 and that there was a 'small...not entirely out of the question possibility' of it hitting the Earth. That prediction should never have been made public, but it did cause quite a stir—the media always looking for a sensational story. NASA, however, was not amused, seeing its credibility as being at stake. And although the MPC proudly carries the IAU name, NASA is, in fact, its major sponsor.

'It was certainly fireworks, and a media uproar without any pre-warning was not to NASA's liking. I went to NASA Headquarters to make amends with Mike A'Hearn as my "bodyguard", and with his reputation we managed to straighten out things. But this made me search the IAU archives for documentation on the relationships of the MPC to the IAU. There was none. A contentious issue was the database of then about 2 million unconnected minor planet observations, which was inaccessible to orbit calculators elsewhere in the astronomical community. My own opinion was and is that if the IAU name is on a database, it must be available to all members of the scientific community, so it was clear to me that the MPC needed proper Terms of Reference, although the MPC leadership saw this differently.

Luckily, Irwin Shapiro, the Director of the Center for Astrophysics (CfA) at Harvard, which hosted the MPC, agreed with me, and the ToR were finally approved in 2002.'

But the MPC would soon again cause newspaper headlines and public commotion. In January 1999, Brian Marsden, who had valid scientific reasons to consider Pluto an asteroid, proposed to assign it an 'honorary' Minor Planet number

(Continued)

(No. 10,000). This was fed to the press and resulted in no less than 900 e-mails of protest to the IAU, especially from American school children. It stopped from one day to the next when I issued a press release saying that there was no proposal at the IAU to change the status of Pluto, and there was no intention to do so until there was a clear definition of a planet. This put an end to the 1999 'Pluto affair'. But as we know, it resurfaced with full force in 2006.

'Brian Marsden, who passed away in 2010, was a brilliant scientist and a fabulous organiser, but also an opinionated and quite colourful person.'

But the IAU also has a public role to play and that this had many facets.

In connection with the celebration of the Ahmed al-Fergani 1250th anniversary in 1998, GS Andersen visited the Ulugh Beg Astronomical Institute in Uzbekistan. A CCD camera resulted. Seen here from left are Boris Artamonov, GS Johannes Andersen, Prof. Shuhrat Ehgambardiev (Director of the Ulugh Beg Astronomical Institute), Mansur Ibrahimov (astronomer at UBAI). Photo: Collection J. Andersen

'The thing I was most proud of on behalf of the IAU during my tenure was the ISYAs, the International Schools for Young Astronomers. There's a picture from an ISYA in Teheran, where the female students in their black scarves are "fighting" for the control of the Solar Telescope and I thought: That's the essence of what the IAU should be doing! However, looking at it as a General Secretary, we already had Commission 46 (teaching of astronomy), the TAD (Teaching for Astronomy Development) programme and the ISYAs, a whole Commission 38 for the exchange of astronomers, and an EC Working Group on the worldwide development of astronomy—all working on what at least I saw as the same

(Continued)

topic. So I thought this ought to be streamlined, put more money into education, and even started a "cook-book" on astronomy development.

If we could build on activities such as those in Vietnam, e.g. twinning astronomy teachers and helping the production of astronomy books for teaching purposes, etc., this could develop into a long-term, much-expanded mission. But then I got caught up in the operations leading up to the General Assembly, and aside from the formal reorganisation, little happened for a long time: I did not think far enough ahead to realise that you need a professional effort and have to find the necessary funds. George Miley has now made it sustainable activity and for that, the IAU—and I—should be grateful. It will be another major raison d'être for the IAU for the next 100 years.'

As regards public visibility, the year 2000 saw the emergence of a new opportunity for the IAU.

'Peter Gruber was a Hungarian-American Jewish philanthropist, who was brought up in India by Jesuits when the family sought refuge in the USA from Nazi persecution. In 1993, he and his later wife Patricia established a charitable foundation, and a few years later they had the idea of creating a Cosmology Prize. For that purpose, they sought the cooperation of the IAU. They had hired an experienced assistant and theologian, Larry Tise, to run the prize programme for them.

The apparent mix of science and religion did not find takers at the IAU Executive Committee, which rejected the proposal for cooperation along the lines originally suggested. However, in an e-mail from the Foundation, they asked if it was still possible to re-discuss the matter.

While understanding the concerns of the Executive Committee, I had this distinct feeling that "here's good will going to waste", so in June 2000 I went to New York to discuss the matter with the Grubers and Larry Tise. I made it clear that the IAU could not take a backseat to anyone in scientific matters concerning astronomy, so part of the solution would be that the IAU was given an adequate representation on the Foundation Board. In addition, I thought that if given a choice, I'd rather spend the money on young and upcoming people than on well-established scientists with long track records of past achievements, so I argued vigorously for creating a fellowship programme instead. After a full day of negotiations in the magnificent Gruber apartment overlooking Central Park, we reached an agreement not only regarding the Cosmology Prize, which they insisted on keeping, but also to supplement it with funding for a fellowship programme. And this agreement was indeed approved by the Executive Committee just before the Manchester General Assembly.'

(Continued)

Since the Manchester GA, the Gruber Prize Award Ceremony has been a firm part of the GA Opening Ceremony. Here, Patricia Gruber is seen with Mandolesi Nazzareno of the Planck Team, one of the 2018 prize winners. Photo: C. Madsen

'The Manchester GA itself is not a good memory to me, despite the sterling efforts of the late organisers, Rod Davies and Dennis Walsh. Mainstream UK astronomy, or significant sectors of it, seemed not to be engaged. But first, to my surprise, there seemed not to be a single page in Paris on how to organise a GA, although it is the biggest expense of the triennium. Then, we tried to make the GA low-cost by having the University offer the lecture rooms for free. This meant a lot of walking between buildings, but you don't get the services of an expensive congress centre at youth hostel rates, and the experiment was never repeated. The "professional" conference organisers were not familiar with the Web, etc. At least I wrote down what I had done and when, for the benefit of my successors.'

After the end of his term, he undertook a revision of the IAU by-laws and statutes as well as the structure of the divisions together with Bob Williams, then an IAU Vice President.

'The General Secretary is the only person who knows every detail of the by-laws and statutes by heart. But when he is in office, he has no time to change them. So now we set out to simplify the documents and make them a tool for change, not an obstacle. For example, Divisions could create or terminate working groups or Commissions. And the new documents were approved by the 2003 XXVth General Assembly in Sydney.'

Andersen's scientific life has focussed on observational studies of stellar evolution, the chemical and dynamical evolution of the Galaxy, and instrumentation. He has served in many forward-looking functions in astronomy, including Director of the Astronomical Observatory of Copenhagen University, Chairman of ESO's Scientific Technical Committee, Director of the Nordic Optical Telescope

(Continued)

(NOT), and Chairman of the Boards of the EU-sponsored networks OPTICON and ASTRONET. Aside, of course, from his services to the IAU. How does he think about his involvement with the IAU?

'When I accepted to become Assistant General Secretary and subsequently General Secretary, I knew I would put 100% of my working time into the IAU and to the best of my abilities, I did. At the same time, in 1994, I became in charge of the first Instrument Centre of the Danish Research Council, the fulfilment of another dream. This would have been great, had the person who was supposed to manage the daily operations not sadly fallen ill and shortly after died. He was truly irreplaceable, and a considerable amount of unforeseen work also ended on my shoulders. So it was a tough time, but I enjoyed every moment of it. Fortunately, I'm married to another astronomer, also with close links to the IAU ...'

Boris Shustov: A View from Moscow

Vice-President (2015–2021)

Interview conducted in Vienna on 1 February 2018

Many IAU officers can look back at decades of close relations with the IAU. Not so in the case of Boris Shustov, who was elected as IAU Vice President in 2015. Yet Shustov's first encounter with the IAU was as an unofficial assistant to Alexander Boyarchuk when in 1991 he became President of the IAU amid dramatic changes in Eastern Europe and Russia.

Boris Shustov in front of the Vienna International Centre, the home of the United Nations in Vienna, including the UN Committee for the Peaceful Uses of Outer Space. Photo: C. Madsen

(Continued)

'During my young years, the Soviet Union might not have been wealthy, but it could still do interesting things. First was Sputnik. At the time, I was a schoolboy in form 7 and I was intrigued by the fact that Sputnik could travel around the Earth and not simply fall down. Then came the mission of Yuri Gagarin, the first human in space. Instantly, most young Russian boys began dreaming of becoming cosmonauts. I simply decided to study space. At the time, I lived in what was then called Sverdlovsk but has now resumed its old name of Jekatarinenburg. Together with other youngsters, I began participating in observations of satellites by simple visual observations. The Astronomical Council of the Academy of Sciences of the Soviet Union had set up a network of satellite trackers in about 60 locations, an important undertaking at the time. Some spacecraft were lost because of imprecise ephemerides and thus the narrow uplink radio beam for sending command signals just missed them. I also constructed my own telescope using lenses from normal eyeglasses mounted on a wooden stick—it was sufficient to look for double stars.

In 1964 I joined the Ural State University as a student of Prof. Klaudia Barkhatova, who had arrived from Moscow. I continued under Dr. Leonid Snezhko, who taught the subject of stellar evolution, an emerging field at the time. After that, I moved to Moscow to work with Prof. Alla Massevitch, who was deputy head of the Astronomical Council, which in reality was a research centre. In Moscow, I was interested in the work of Bohdan Paczyński and Rudolf Kippenhahn. This way, we became exposed to new technologies and computers and it helped us enormously. During that time, I took the subject of star formation to my heart. Still, in 1971, for personal—mostly financial—reasons I took a detour as a computer scientist at Zvenigorod satellite tracking station of the Astronomical Council. This brought me to work at satellite tracking stations across the world.

Alla Massevich at the New Delhi GA. Photo: AIP Emilio Segrè Visual Archives, John Irwin Slide Collection

(Continued)

Then, in 1987, Alexander Boyarchuk came from Crimea Observatory, at the time the no. 1 facility in the Soviet Union, to take over as President of the Astronomical Council. Boyarchuk expanded the activities of the Council by adding space astronomy. He was heavily involved in the "Astron" project, the Soviet 80 cm UV Space Telescope (named Spica) that was operational between 1983 and 1989. I joined his team, and thus life—fate, perhaps—has enabled me to look at different areas such as astrophysics, satellite science, space astronomy and technology.

In 1988, Harry van der Laan invited me to come to ESO for half a year. I also became better acquainted with Western astronomy, even becoming Vice President of the European Astronomical Society between 1993 and 2000.'

Working with Boyarchuk, Shustov was involved with ambitious space astronomy projects, including the 'T-170', a 1.7-m UV telescope follow-up to the Spica Telescope onboard the 'Astron' mission. But then, the political earthquake in 1991 shattered all plans. Instead, a dark age befell Russian and Central European astronomy in the wake of the collapse of the Soviet Union.

'It was a hard time. I remember one observatory that had electrical power for one hour only each day—during the daytime, that is. Luckily, we received help from our Western colleagues and institutes, both financially and equipment-wise. Among others, ESO stepped in but was subsequently replaced by the EU-supported INTAS scheme. It should be said explicitly that Russian astronomers remain thankful to their colleagues from Europe, the United States and elsewhere for the help they gave us during that difficult period.'

Around the same time, at the 1991 IAU General Assembly in Buenos Aires, Boyarchuk became elected as President of the IAU.

'Boyarchuk's English was rather poor, so he asked me if I would be willing to help him during his term as IAU President, e.g. by translating letters and other texts.

As already mentioned, this took place at a time of great societal upheaval in Eastern Europe and Russia. In any event, Boyarchuk undertook great efforts to ensure that all the countries that had been part of the Soviet Union received help and assistance on equal terms and to support the links between these countries and the IAU.

When Boyarchuk took the helm of the IAU, calls for reform of the Commission and the very structure had become loud. Both the outgoing and incoming General Secretaries supported it and quite some preparatory work had been undertaken. But Boyarchuk was cautious in relation to radical change. As a Ukrainian with intimate knowledge of the precarious situation of science—and scientists—in the fSU (the former Soviet Union) and Eastern Europe. Boyarchuk cared about preserving the consent and retaining the national support for the IAU.

I respected Boyarchuk for his wisdom, but it's correct to say that he was widely seen to be rather conservative. If someone proposed changes, he wanted to be convinced, and given the fact that a number of Commission presidents had issues with the proposed changes, he, too, was not convinced.

(Continued)

Alexander Boyarchuk is seen here at the centre of the photo to the left of Monique Orine from the IAU Secretariat. Other people in this photo that was obtained during the EC68 meeting in Baltimore in early 1996: (From left to right) Bambang Hidayat, Jozef Smak, Bob Kraft, Elaine Williams, and Virginia Trimble. Photo: J. Andersen

From the 1991 General Assembly, I especially remember the resolution on hazardous Near-Earth Objects. It was not part of my scientific interests at the time, but then in August 1999, 1999 JM8, a 7-km large body passed the Earth at a distance 8.5 million km. It sharpened my attention to the problems that such objects can cause to us. Indeed, with the largest landmass on Planet Earth, Russia is, of course, especially at risk—a fact that was driven home with the Chelyabinsk incident on 13 February 2013. In any event, the NEO challenge came back to me, when, in 2007 and now as a member of the Russian Academy for Science, I was charged with setting up an expert group on space threats, that also brought me in contact with the UN framework regarding the Near-Earth Objects, UNCOPUOS.

Given my history, I was taken by surprise by the invitation to become a candidate for a Vice-President of the IAU, by the special nominating committee.

As IAU Vice President, my area of responsibility comprises our links with other international bodies, such as ICSU, ITU, UNCOPUOS, COSPAR, etc. This has over time at least occasionally been a somewhat neglected area, but, in fact, we have links with about 30 bodies. It's a new area for me, but I am determined to do my best to establish a more formal system and thus support these activities.'

5

Opening Up (1990–2000)

Initial Optimism

The fall of the Berlin Wall in 1989, the launch of the Hubble Space Telescope in 1990 and the opening of the first 10-metre sized telescope in the same year (Keck I) started a decade of optimism, both in astronomy and in international cooperation. The end of the Cold War, the end of Apartheid in South Africa soon after, and the intensification of European cooperation within the European Union, opened up a lot of new opportunities for international cooperation. For the first time since the foundation of the IAU, international relations were not dominated by a major geopolitical conflict. Economically, the end of Communism was difficult in Eastern Europe and the former Soviet Union, which also affected astronomers, but the atmosphere of internationalism combined with economic growth in the West made it relatively easy to obtain support for new international projects.

For the IAU, this period could be characterised as one in which the Union opened up within the astronomical community. Many new countries joined, and there was increasing attention to the role of women (by now around 10% of membership) and early career researchers. The General Assemblies were reorganised to increase their scientific appeal, and finally a beginning was made with the reorganisation of the internal structure of the Union, with the introduction of Divisions in 1994. A small but significant new initiative concerned the Teaching Astronomy for Development (TAD) programme.

At the same time, the IAU was confronted with several public debates, forcing it to adopt a more public role in issues such as the potential hazard of Near Earth Objects and the status of Pluto. Some of these issues would

© Springer Nature Switzerland AG 2019
J. Andersen et al., *The International Astronomical Union*,
https://doi.org/10.1007/978-3-319-96965-7_5

become much more urgent after the turn of the century, but in the 1990s, the first signs of the new Union had already become visible.

The IAU extended several international support and exchange programmes to specifically support former Eastern astronomers, in part funded by two grants from NASA (via the AAS).[1] Many new nations could join the IAU in rapid succession. Following ICSU recommendations, former Soviet Union republics were welcomed automatically, as were Slovakia and the Czech Republic. The fact that the IAU President at the time was a Russian, Alexander Boyarchuk, undoubtedly helped. The number of members from Asia, South America and the Middle East also increased. A group of five Central American countries joined collectively in 1997 as the Central American Assembly of Astronomers (CAAA).[2] This example of collective membership was later abolished as unsuccessful, so it remained a unique case.

This did not mean that political issues were a thing of the past completely, however. For the IAU, the two most important issues concerned China and the former Yugoslavia. While the breaking up of the Soviet Union and Czechoslovakia was mostly peaceful, Yugoslavia descended into war in the early 1990s. The UN, released from the Cold War gridlock, called for international sanctions against Yugoslavia (i.e. the remaining part, dominated by Serbia). ICSU followed, and therefore also the IAU; Yugoslavian membership was terminated, and new Serbian membership would have to wait until it had been accepted as a member by ICSU.[3] The international sanctions also meant that Serbian astronomers were denied entry to the Netherlands for the 1994 General Assembly, despite protests from the organisers; Serbia finally rejoined the Union in 2003. As in the previous decades, visa trouble remained the most common effect of the political tensions—Chinese astronomers also had difficulties entering the USA, for example.

While communist regimes were overthrown one after another in Europe, demonstrations on the Tiananmen Square in Beijing were suppressed violently in 1989, causing strong international protests. For the IAU, this was sensitive, since China had only re-joined a few years earlier, and the IAU Executive Committee included a Chinese Vice President. At stake were two colloquia planned to take place in the country that summer. Moreover, a prominent Chinese astronomer and dissident had sought refuge in an embassy

---

[1] IB 68 and IB 70; see also correspondence in IAU Archives box 18f, file 'USSR'.

[2] IAU Archives box 1B, file 'CAAA'. Its membership was terminated in 2002 because of arrears in paying the membership fee.

[3] EC minutes 65 (1994); additional documents in IAU Archives box 1A.

in Beijing and asked for free passage to leave the country. Once again, steering clear of politics was difficult: cancelling would be as problematic as the opposite. After painful discussions, the meetings were 'postponed' because there was 'insufficient interest' from outside China to make the meetings 'truly international'—a diplomatic formulation, similar to the one used in 1951 when the General Assembly in Leningrad was cancelled. The Executive Committee emphasised that this decision was made by the scientific organisers; and the dissident astronomer's problem found a satisfactory solution through the normal diplomatic channels.[4]

Despite these setbacks, astronomy in China continued to develop with astonishing speed. Other countries that quickly rose to prominence in astronomy were Spain, especially after the opening of the Canary Islands observatories, and South Africa, where the post-Apartheid regime invested heavily in astronomy, which became a symbol of both economic and cultural development.

A New Telescope Boom

Astronomers had dreamt of space telescopes since the start of spaceflight. Since the 1950s, satellites had observed radiation in wavelengths that were blocked by the Earth's atmosphere, but for optical light too, going to space would be beneficial: in space, a telescope would be free from the atmospheric turbulence that limits the resolution of observations from the ground. In the 1970s, the Space Telescope project started in earnest, designed for both optical and near-UV observations. The project was run by NASA, but it also had an international component, including a 15% participation by the European Space Agency.[5]

But humans plan, and events decide. After major delays and significant cost overruns, the Hubble Space Telescope, as is was now called, was finally launched in 1990, only to find that its primary mirror suffered from spherical aberration. The telescope was different from earlier space observatories, however, because it had been designed to be serviced by astronauts: its instruments could be renewed. This made it possible to install corrective optics in 1993. Ever since, the Hubble Space Telescope has delivered an endless succession of scientific discoveries and spectacular images, and today it is hard to

---

[4] IB 64 (1990).
[5] Smith (1989).

imagine any astrophysical field that has not been profoundly changed by it. The images also captured the public imagination; several photos released by the Space Telescope Science Institute have become cultural icons. When NASA planned to cancel the last service mission to the telescope after the Space Shuttle Columbia disaster in 2003, it had to reverse that decision under severe public pressure.

The Hubble Space Telescope photographed during the second servicing mission in 1997. Photo: NASA

The Hubble Space Telescope was the most visible of the new generation of instruments that also included new infrared, X-ray and gamma ray space telescopes. The 1990s also finally witnessed the opening of a new class of ground-based telescopes of 8–10 metres, starting with the Keck I Telescope in Hawai'i and culminating in the European Very Large Telescope that consists of four 8-metre telescopes, which were taken into service around 2000. Other notable telescopes include the Gemini Telescopes, the Subaru Telescope, the Gran Telescopio Canarias, the Southern African Large Telescope and the Large Binocular Telescope. Most of these telescopes were developed by national or international consortia. Apart from the increased size of the telescopes, the quality of observations also improved with the introduction of active and adaptive optics (electronic control of the mirrors to counteract slower or more rapid distortions caused by the optics or the atmosphere).

Observing time was allocated by competitive selection of proposals. As big telescopes were rare, the number of astronomers grew, and the physics more complex, more and more research was done in teams rather than by individual researchers. The percentage of articles by multiple authors rather than single-author papers rose significantly.[6] This happened later in astronomy than in many other natural sciences.

Apart from general-purpose telescopes there were also more specialised instruments, such as the COBE satellite (launched 1989) that mapped the Cosmic Background Radiation, and the HIPPARCOS satellite (also 1989) that measured the position of stars with unprecedented precision. Early measurements of the cosmic microwave background (CMB) radiation did not cover the entire range of wavelengths or frequencies. One of the problems was the need for long, uninterrupted observation series, which could only be done from the Antarctic plateau or in space. The eagerly-awaited results of the COBE satellite caused the audience to break into spontaneous applause: an absolutely perfect blackbody curve for a temperature of 2.7 Kelvin. COBE also measured the 'anisotropy' of the background radiation: very small fluctuations that give an indication of the characteristics of the early universe. The basic nature of the Cosmic Background Radiation and the fact that it indeed is a relic from the Big Bang were now firmly established, and the principal investigators, John C. Mather and George F. Smoot, were awarded the 2006 Nobel Prize in physics for this achievement. Later, several space missions and ground-based experiments have analysed the background radiation in ever greater detail, including its spectrum of spatial frequencies (WMAP, 2001) and its polarisation (Planck, 2009).

The HIPPARCOS mission marked the revival of astrometry, the classic art of measuring stellar positions, which had been overshadowed by astrophysics in the twentieth century. This time, it was done from space, unhindered by turbulence in the Earth's atmosphere. The satellite observatory was launched by ESA in 1989 and operated until 1993. Due to a technical problem, it never reached the intended geostationary orbit, causing some frantic reprogramming. It proved possible to still reach the intended accuracy, however, and the final catalogue contains position and parallax of nearly 120,000 stars to extremely high accuracy (0.001 arcsec), and many more with slightly lower (but still high!) precision. Moreover, the stability and homogeneity of the

---

[6] Heck vol I (2000).

HIPPARCOS data are such that the fundamental (inertial) International Celestial Reference System (ICRS) is now based on HIPPARCOS. Endorsing the new inertial reference frame is a classic task for the IAU; the new reference frame will eventually be replaced by the even more accurate results of the on-going Gaia mission around 2020.

Artist's impression of the Gaia launch into orbit. Photo: ESA/D. Ducros

All this resulted in an impressive amount of new data—in electronic form, since photographic technology had by now been completely replaced by electronic detectors. The digital data could be exchanged through the rapidly expanding Internet, and analysed with ever more powerful computers. With time, the computers became so fast and powerful that 'computational astronomy', based on detailed numerical modelling of physical processes, became a specialised branch of astronomy in itself, and a new source of knowledge next to observational and theoretical astrophysics.

The wealth of data and increasingly detailed computations brought many new insights in the physics of the Universe. High-energy astrophysics advanced rapidly, as did the theoretical understanding of the interstellar matter, which turned out to be surprisingly rich in complex molecules. The new observations also raised new questions, however, especially after the discovery in 1998 that the expansion of the universe is accelerating. This was the opposite of what everybody had expected and discussed: a uniform or even slowly decelerating expansion. The evidence came from Type Ia supernovae, which had been discovered to larger and larger distances, thanks to modern technology and more efficient search strategies. Moreover, the correlation between their peak luminosity and their decay times made it possible to use them as 'standard candles'.

Two competing teams had used different techniques to reach the same result, and their papers announcing the discovery were published nearly simultaneously. Saul Perlmutter, Brian Schmidt and Adam Riess also shared the 2011 Nobel Prize as well as the Gruber Cosmology Prize for this discovery. Their discovery supported cosmological models that stated that visible, baryonic matter only encompasses a mere ~5% of the Universe; later modelling concluded that the other 95% consists of 'dark matter' (~25%), the existence of which had already been concluded from other observations, and an even more mysterious 'dark energy' (~70%), the nature of which is still completely unknown.

Interestingly, one of the most spectacular discoveries of the 1990s was done with a relatively small telescope. In 1995, Swiss astronomers observing at the classic Haute-Provence observatory in France announced that they had found a planet with a surprisingly short period of only 4 days, orbiting another star than the Sun. There had been earlier claims, but this was trustworthy enough to inaugurate a rush to find new 'exoplanets', as they were called.

Earlier attempts had all assumed that exoplanetary systems were analogous to the Solar System, but the first actual exoplanets were gas giants in extremely close orbit around their star. They did not resemble anything in the Solar System at all, so all models could be thrown out the window. These massive, short-period 'hot Jupiters', as they were soon dubbed, were of course the easiest to discover, but when transiting exoplanets were also found, giving access to approximate radii and densities, sceptics were silenced. Overnight, exoplanet research became a major field of astronomy.

The desire to understand exoplanetary systems in general is also motivated by the fundamental, underlying human question: are we alone? Does life exist elsewhere in the Universe, and if so, what form(s) could it take? The recent Kepler space mission (launched 2009) has shown that exoplanets are in fact very common—at least as common as single stars—and exist in bewildering variety. That is still several steps short of finding life, but our own solar system turned out to be less unique than many people had tacitly assumed. It is no coincidence that a prominent goal of all the future giant optical telescopes now under construction is the direct observation of exoplanets.

IAU: The Beginnings of Change

As we have seen several times before, the role of the IAU in international infrastructure collaborations was ambiguous. Because it lacked its own financial clout, it could not propose or manage instruments, but that also meant that it could provide a neutral ground. And as instruments got bigger and more expensive, and therefore rarer, global coordination was clearly desirable. In 1994, the General Assembly featured an extra meeting on 'Future Large Scale Facilities in Astronomy', and a working group was founded for that topic. Ten years later, however, it reported that activities in this field were 'becoming inherently international without external IAU-led promotion'.[7] Still, the IAU provided a framework for international meetings—certainly on science, if not on instruments.

The number of symposia and colloquia kept growing. The General Assemblies were still major international events, but as we saw in the previous chapter, there was also much discussion on their structure. In the 1990s, this finally led to change.

Increasingly, the GAs attracted participants from new countries outside the sphere of traditional Western astronomy. This picture is from the 1991 GA in Buenos Aires. Photo: IAU Archives

The 1991 General Assembly in Buenos Aires, the first in South America, still followed the old format, although poster sessions were added to the

---

[7] IB 96 (2005) 10.

programme to attract younger researchers (this was described in the IAU Bulletin as 'a major departure for the Union'![8]). At this occasion, the General Assembly voted to name the 5000th minor planet after the IAU. The preparation for the General Assembly had been somewhat turbulent because of Argentinian internal politics and the high inflation, but most participants will remember the meeting mostly for the fire that broke out in the car park under the conference centre in the early morning of the final day, making the building unusable. With a lot of improvisation, most sessions could be saved, including one on early Hubble Space Telescope results.

Incoming IAU President Lo Woltjer, General Secretary Jacqueline Bergeron, and the organisers of the 1994 General Assembly in The Hague initiated a new format. The number of administrative (commission) meetings was limited, and all the symposia from the General Assembly year were included in the General Assembly, rather than being organised just before or after it, as had been common. This offered people the opportunity to attend more scientific meetings of their own choice for the same investment in travel funds and—more importantly—time. In this way the scientific appeal of the General Assembly was strengthened. The meeting was still long, but this conference was now supported by both ESO and ESA. The new format was received positively, although some missed the somewhat chaotic 'bazaar'-like atmosphere of previous meetings.[9]

---

[8] IB 63 (1990).

[9] Correspondence in IAU Archives box 6, file 'Format General Assemblies 1990–2006'.

Ed van den Heuvel, Chairman, Netherlands National Committee for Astronomy (NCA) welcomes Queen Beatrix at the 'Congresgebouw' of The Hague. Photo: Collection Ed van den Heuvel

That IAU General Assemblies are major events is also illustrated by the fact that many heads of state have attended its openings, including Prime Minister Rajiv Gandhi in Delhi, President Carlos Menem in Buenos Aires and Queen Beatrix in The Hague. Special highlights were the presence of the Imperial Couple of Japan in Kyoto in 1997, which had been prepared in the greatest secrecy, and in 2012 in Beijing of then Vice-President, now President of China, Xi Jinping.

Members and Divisions

As we have seen above, many new countries were eager to join the IAU, boosting the number of both national and individual members (see the list of member countries and membership numbers in Appendix A). The diversity of the membership changed only slowly, however. In 1993, only 28% of the members were less than 45 years old; almost 40% were between 45 and 55.[10] The percentage of women was only just above 10%, although this varied much between member countries; in 2001 it had risen to 12%. The Algerian and Peruvian delegations were 100% female, but both consisted of only one person. Of the larger countries, France scored high, with 26% women among its members, while neighbouring Germany had less than 4% women delegates, Japan even less.[11] In general, Latin American, Southern European and Eastern European countries had relatively high percentages of women. In most countries, the share of women among IAU members was lower than in the astronomical community in general, indicating that they were proportionally even more under-represented among senior researchers.[12]

The underrepresentation of women in the IAU finally became a topic of explicit discussion in the IAU. This was not self-evident: even after the Baltimore meeting of 1988 (see previous chapter), in 1991 a panel discussion during the General Assembly in Buenos Aires had to be scheduled outside the official programme, as General Secretary Derek McNally judged the topic too 'political'. Several incidents in 1991–1992 made the issue more urgent. While the incidents themselves were minor—some remarks inadvertently causing offence—they led to much debate, highlighting a more structural issue. For example, women were underrepresented in invited lectures, even compared to the percentage of the membership. A symposium in Australia in 1990 had only one woman among 56 speakers, and only two of the 15 invited discourses at General Assemblies from 1973 to 1991 had been delivered by a woman (Vera Rubin and Alla Massevitch in 1985, the latter being a last-minute replacement for Roald Sagdeev).

---

[10] Membership statistics by age in IB 70.

[11] Débarbat (2004); see also membership statistics in IB 68.

[12] Cesarsky (2010) and Débarbat (2004).

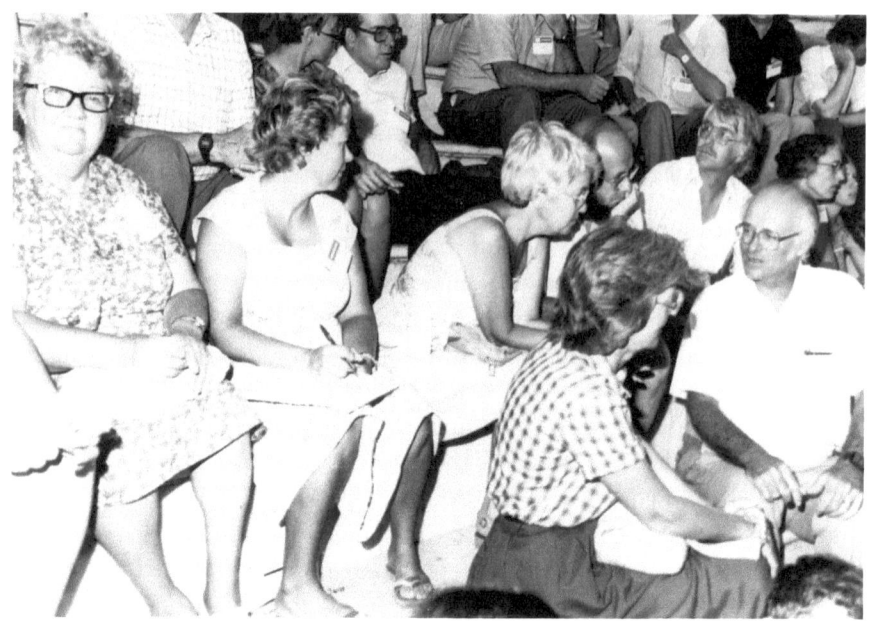

Vera Rubin (leaning forward at the centre), Bob Rubin (hand on chin), Jan Burke (plaid blouse) and Bernard F. Burke (bottom right) at the IAU GA in Patras in 1982. Photo: AIP Emilio Segrè Visual Archives, Rubin Collection

After these issues were publicly addressed by several women, new formal guidelines for IAU meetings were first formulated in 1997. Meetings on the role of women became standard features in the programme of General Assemblies. From now on, the topic was firmly on the agenda.[13]

---

[13] As witnessed by documents and correspondence in IAU Archives box 34 'Women in Astronomy'.

Women in astronomy luncheon at the GA in Rio de Janeiro. This has evolved to become a regular feature of the General Assemblies with ever increasing participation. Photo: IAU Archives

McNally once defined the mission of the IAU as: 'The IAU must provide the glue that holds the astronomical community together internationally.'[14] This can of course be interpreted in various ways. In earlier decades, the unity of astronomy had been challenged both by political fault lines, and by the growth of subdisciplines that threatened to lose connection with the rest of the discipline. In the 1990s, these threats were no longer urgent. But the large number of Commissions was often mentioned as a problem. What if researchers with closely related interests missed each other because they were in separate Commissions? The commission structure was increasingly regarded as outdated. Some were defined by topic, some by method or technology, and some had very narrowly defined topics, all of which led to fragmentation. Even more importantly, some newer topics were not covered by any commission.

Changing the commission structure was not easy, as we have seen in previous chapters. All attempts to change the commission structure had met with strong opposition. This could be interpreted as conservatism of (mostly elder) commission officials, but one could also understand it as a sign that many astronomers apparently found the existing Commissions relevant and important. They identified with them. Besides, being President of an IAU

---

[14] McNally to Pecker, 26 February 1991, in IAU Archives box 6, file 'Format General Assemblies 1990–2006'.

commission was considered prestigious, so both current and potentially future presidents had a strong interest in maintaining existing Commissions. This was not just because of their own ego (though that undoubtedly played a role sometimes), but it could also be important for career opportunities and even to get political support for astronomical projects and institutes. The IAU's seal of approval carried some weight.[15]

IAU President-Elect Woltjer and new General Secretary Bergeron decided to put commission reform on hold for the moment, and start reforming the IAU structure in a different way: by grouping Commissions together in eleven (later twelve) thematic Divisions.[16] Something similar had actually been proposed in the 1960s by Donald Sadler (see Chap. 3). The Divisions were designed to have roughly similar sizes in terms of membership, meaning that some consisted of only one large Commission. The division boards could endorse meeting proposals by 'their' Commissions and organise joint discussions, stimulating more interaction between the Commissions, and hopefully encouraging restructuring in the future. The division presidents acted as advisors to the Executive Board. The creation of the Divisions was endorsed by the General Assembly in 1994, and the new structure was fully implemented in 1997. Since the Commissions remained unchanged, including their names and numbering, the change was relatively small in practice, however. This only changed with a renewed, more fundamental reorganisation in 2012–2015 (see next chapter).

Education and Development

The educational activities of the IAU continued in the 1990s, with the International Schools for Young Astronomers as the main activity. ISYAs were held in Morocco, China, India, Egypt, Brazil, Iran, Romania and Thailand, each with around 40 participants (see Appendix D).[17] From 1992 to 1997 the coordinator of the programme was Donat Wentzel (USA), with Michèle Gerbaldi (France) as his deputy and successor, assisted by Edward Guinan (USA) and then Jean-Pierre de Grève (Belgium).

---

[15] Cf Bambang Hidayat, AIP interview (1997).

[16] Proposal by Woltjer, 13 August 1994, in IAU Archives box 14; it was discussed in several EC meetings.

[17] https://www.iau.org/science/grants_prizes/iau_grants/international_school/list/

For countries with a national astronomy programme, part of the school could take place at an observatory site for training in observational techniques. However, if this was hard to organise, then observational training and data reduction could be provided through the use of remote robotic telescopes. The range of possible activities at the Schools also changed with the increasing availability of digital data archives (both research data and literature) that could be accessed remotely. Activities during current ISYAs have become increasingly computer-oriented, both as regards querying the relevant databases and developing the participants' skills in database or archive 'mining', an e-science concept that has gained significant importance during the last 25 years. It was now possible to do high-level research without privileged access to a large telescope.

There were also other initiatives, aimed at fostering professional astronomical activities in developing countries. In 1994, Don Wentzel initiated a new teaching programme, not aimed at astronomy students such as the ISYA, but at creating small cores of astronomy teachers at all levels, including secondary schools: the Teaching for Astronomy Development (TAD) programme. It typically comprised lectures in basic modern astronomy for 20–30 present and future science teachers in developing countries, with the goal to assist countries that wish to improve the overall environment for science education and research in this way. The TAD programme format was designed to be flexible; its philosophy was to be pragmatic and support the end goal in whichever way was judged the most effective. The activities undertaken depended entirely on local circumstances.

The first TAD programme was organised in Vietnam, where almost all scientific activities had been interrupted for reasons all too well-known. A two-year effort was initiated in 1997, focussing on a summer school to re-introduce astronomy to the country after the 30-year hiatus. A workshop then took place at Vinh University in Vietnam for university instructors and physics students, to update them in modern astronomy, and develop hands-on astronomical activities. Since then, and regularly until 2007, workshops and lectures have taken place in the Vietnamese universities at Vinh and Hanoi. A first bilingual modern textbook in astrophysics (Vietnamese and English) was also written, with colour illustrations, for an astronomy course taught at ten pedagogical universities in Vietnam. It was published in 2000 with partial financial support by the IAU TAD programme. Other (shorter-term) TAD programmes were organised in, for example, Morocco and the Philippines.

For countries in which even a TAD was too ambitious to start with, there was the *Working Group on the World-Wide Development of Astronomy*, established in 1987 as the working group on 'Promotion and Development of

Astronomy' at the initiative of President Jorge Sahade, under the aegis of the Executive Committee. It was chaired initially by Alan Batten and advised non-member countries on how to develop astronomy.[18] Its aim was to initiate early astronomical activities in countries which had none, but where a flame might be lit that could eventually develop into more substantial activities. The scheme was not very effective in practice, however, despite the efforts of a series of enthusiastic Working Group Chairs, especially since it proved hard to organise a long-term follow-up.

Ideally, all these programmes would work together: the working group on Worldwide Development of Astronomy could help prepare the way for a TAD, which could be followed by an ISYA, possibly leading to full IAU membership. In practice, the success of the various schemes varied (especially the visiting lecturer programme remained difficult). At the same time, the activity level of Commission 38 was declining, in part as a result of the appearance of several alternative funding sources for similar purposes (exchange programmes), which suggested that the funds might more profitably be reallocated elsewhere. Besides, although the proliferation of initiatives indicated broad support, it also brought the danger of fragmentation.

In summary, the situation invited a more comprehensive and systematic 'end-to-end' approach to the overall concept of 'capacity building', which this ensemble of initiatives was increasingly called (and 'TAD' was redefined to mean 'Teaching Astronomy *for* Development'—which was its true intention). Accordingly, in 2000 these previously separate IAU activities were grouped together in a re-organised Commission 46, with a 'Program Group' responsible for organising each activity, and an (increased) common budget to be shared flexibly between the different programmes as circumstances dictated. But it placed a lot of responsibility on the Commission Presidents' shoulders, and they changed every three years.

Representing Astronomy

Throughout the twentieth century, the IAU was first and foremost an organisation of astronomers, for astronomers. It had formal contacts with other scientific organisations, but it was probably not very well known among the general public. This started to change in the 1990s, as the IAU adopted a new role: representing astronomy in public. A small, but significant first step was

---

[18] IB 63 (1991); also Alan Batten (S349 forthcoming).

the creation of the IAU's very first website, personally designed by General Secretary Appenzeller around 1994 and hosted by the University of Heidelberg. Today's IAU web site is hosted and maintained in collaboration with ESO. But several public issues also required more active interference on behalf of the Union, most notably protection of the night sky, minor planet nomenclature, and the Near Earth Object issue.

Defending the interests of astronomy against light pollution and radio interference was rapidly becoming a key task for the IAU. To highlight the issue, in 1997 the General Assembly declared that the night sky was 'the heritage of all humanity, which therefore should be preserved untouched'. UNESCO's 'World Heritage' programme appeared to be a good 'role model', but turned out to be inapplicable in practice. And the task was not easy: apart from light pollution, pollution of space itself with 'space debris' became a serious issue as well, as the number of satellite launches increased. In 1993, an American firm even proposed to create 'Space Billboards' for commercial advertisements, which when visible would be as bright as the full moon. Luckily, the project never materialised, for practical and financial reasons.

In 1992, the IAU, COSPAR (Space Science), ICSU and UNESCO organised a joint meeting on Adverse Environmental Impacts on Astronomy in Paris. The issue was also addressed by the UN *Committee on the Peaceful Uses of Outer Space* (UN-COPUOS), for example at the UNISPACE III Conference in 1999, jointly with which the IAU organised a dedicated Symposium (No. 196, *Preserving the Astronomical Sky*).

Protecting radio frequencies also became increasingly problematic, as radio waves were being observed at all wavelengths, and radio telescopes became ever more sensitive, while transmitters were mounted on innumerable moving satellites. Fortunately, optical fibre cables soon offered a much cheaper and more efficient way of transmitting data, outcompeting the expected boom in satellite communications. As the UN *Committee on the Peaceful Uses of Outer Space in Vienna* was the main forum to discuss the political and legal aspects of both protection of the night sky and the hazard of Near Earth Objects (see below), the IAU became more active in that committee, having obtained official status as permanent observer in 1995.

The UN Committee on the Peaceful Uses of Outer Space (UN-COPUOS) in session at the UN Vienna International Centre. Photo: C. Madsen

Defending the interests of astronomy got a new dimension as the IAU was confronted with commercial ventures that claimed to sell names of stars and other astronomical objects. This became a recurrent issue. It touched upon one of the core functions of the IAU: coordinating nomenclature. Real intervention was only possible if the IAU name or logo was abused; the IAU's authority was well-established, but it did not rest on a legal framework. The Union could only protest, or mock the 'enterprises selling fictitious goods', for example on its website.[19]

The IAU's authority on naming of minor planets was uncontested, but here too, some problems arose. For one thing, the number of minor planets grew so rapidly that the Working Group for Small Bodies Nomenclature could hardly keep up. Moreover, they were more and more often approached by people or interest groups with proposals, sometimes offering money for a name. Others wanted to organise public outreach activities with a minor planet name as 'prize'. Besides, some proposed names were published in the media before they were approved. All this meant that increasingly strict procedures had to be put in place to manage the naming of (minor) planets. At

---

[19] see https://www.iau.org/public/themes/buying_star_names/

the same time, the process was also speeded up: from 2006, the approval of new names did not have to wait for the next General Assembly anymore. From now on, names were published continuously. If someone protested within three months, the name could still be changed.[20]

Another issue that attracted attention was the status of Pluto. Since its discovery by Clyde Tombaugh in 1930, it had been counted as the ninth planet of the Solar System, but it seemed like an odd planet compared to the others: it was much smaller, and it had an unusual orbit. Some people, including Brian Marsden of the Minor Planet Center, thought that it looked more like a minor planet. The status of Pluto turned out to be a sensitive public issue, however. In 1998, discussions about this were picked up by the media, obliging the IAU to issue a reassuring press statement that Pluto's status as a planet was not under threat, given that no difference between minor and major planets had been defined. But the issue did not go away, returning with a vengeance in 2006 (more on both the 1998 and the 2006 discussions in the next chapter).

The Near Earth Object Issue

In the twentieth century, humans and human-made probes have visited astronomical objects for the first time, but astronomical objects have also been 'visiting' the Earth for a long time. Most of these objects consisted of dust and small rocks that left no mark, or at most an ephemeral trail of fire as they burned up in the atmosphere. Some had a bigger impact, however.

Old impact craters erode away in a few thousands or millions of years, but those that survive until the present prove that asteroids have impacted in the past, and will do so again in the future. Well-known historical examples are the huge Chicxulub crater in Mexico, which is believed to have caused the demise of the dinosaurs 65 million years ago; the ~50,000 year-old Barringer meteor crater in Arizona; and the Tunguska impact in Siberia in 1908. The public awareness that impacts are not only of historical interest, but could happen at any time, was no doubt greatly spurred by the ringside view of the spectacular impact of Comet Shoemaker-Levy 9 on Jupiter in 1994, after the comet had been gravitationally captured by the planet in a close encounter a few decades earlier. The even more recent Chelyabinsk meteor in 2013 in Russia was another reminder. Although it was only about 17 metres across, it

---

[20] IB 101 (2008) 29.

exploded with the force of a 500-kiloton atomic bomb, injuring hundreds of people and causing considerable material damage on the ground. So the question when and where the next impact may happen, and how big it will be, is an urgent one.

Searching for and calculating, mapping and cataloguing the orbits of potentially hazardous asteroids (PHAs) is one of the activities where astronomy—and more specifically the IAU—can render a real service to society. It is one of these difficult questions, where a natural disaster may cause extremely serious damage—potentially including the extinction of our entire species—but occur at extremely long intervals compared to human lifetimes or even civilisations. In contrast to other, more familiar natural disasters, however, asteroid impacts can in principle be predicted decades in advance, and thus possibly be prevented. Mitigating an actual impact is a job for others, but discovering them and calculating their orbits is clearly a job for astronomers. Cataloguing and keeping track of 'Near Earth Objects' has become a job for the IAU.

Near Earth Objects (NEOs) are minor Solar System bodies (asteroids and comets), the orbits of which may pass that of the Earth within 1.3 AU (Astronomical Units—the mean distance between the Sun and the Earth). Known NEOs include more than 17,000 near-Earth asteroids, while perhaps 100 appear as comets. The great majority of Near-Earth Asteroids normally orbit the Sun near the ecliptic, in the so-called Asteroid Main Belt between the orbits of Mars and Jupiter, but their orbits can be deflected into the inner Solar System through a close passage of a major planet. Most comets approach the Sun only once and may appear at great speed from any direction; a few can likewise pass a major planet on the way through the Solar System and be deflected into an orbit of relatively short period. If a NEO's orbit crosses that of the Earth and the object is larger than 140 metres across, it is called a Potentially Hazardous Object (PHO).

The first and largest asteroid in the Main Belt, Ceres, was discovered in Palermo by Giuseppe Piazzi in 1801, soon followed by three others almost as large (Pallas, Juno, and Vesta). Initially they were known as 'small planets'. After a hiatus of over forty years, asteroid discoveries began to increase. In 1891, Max Wolf pioneered the use of astrophotography and dramatically increased the detection rate of asteroids, which appeared as short streaks on long-exposure photographic plates or as moving objects on pairs of photographs, typically taken on successive nights, and thus were easy to spot by a quick visual inspection.

Today, more than 500,000 known asteroids have multiple observations, and a preliminary orbit has been computed. The Minor Planet Center (MPC) collects all observations of Near Earth Asteroids (including many made by advanced amateurs) and assigns a unique number to each one, so they can be securely identified in the future. The MPC issues regular 'Minor Planet Circulars' with the latest findings. It also keeps track of comets, in close cooperation with the Central Bureau for Astronomical Telegrams (CBAT), which is also located at the Smithsonian Astrophysical Observatory. The Telegram Bureau's task is to provide quick information (initially by telegram) about fast-changing astronomical phenomena; comets are a prime example. The MPC is operated under the auspices of the IAU, although most of the funding comes from other sources (notably from NASA).

The IAU's Working Group on Small Bodies Nomenclature (now within Division F) coordinates the naming of these objects. Only 21,000 (4%) of all the more than 750,000 known asteroids of all sizes have yet been named, however. That fraction will probably not increase significantly in the future.

Brian Marsden at the 1988 General Assembly in Baltimore. Photo: AIP Emilio Segrè Visual Archives, John Irwin Slide Collection

Brian G. Marsden, the legendary Director of both the CBAT (1968–1999) and the MPC (1978–2006), greatly expanded the computational and other services of the MPC. Marsden was a brilliant scientist and organiser, who coped expertly with the steeply increasing influx of new observations and calculated orbits. He was also well known as a vocal participant in any discussion involving small solar system bodies, including the status of Pluto and the issue of NEOs.[21]

---

[21] See for example Brian Marsden, AIP interview (2005).

During most of its existence, the IAU regarded keeping track of NEOs as a routine scientific housekeeping duty. Searches for potentially hazardous asteroids ramped up, however, as the hazard of a future impact began to be appreciated in the 1980s—and as systematic surveys to discover them in time to react became more feasible. This was also a time when sensitive CCD detectors began to replace photographic plates also in wide-field telescopes, and computational means and techniques made both calculation of orbits and matching them with new and earlier observations much more efficient.

The danger of Near Earth Objects started to attract public and political attention after the end of the Cold War (perhaps because the more immediate threat to human existence seemed to have receded for the time being?). At the General Assembly in 1991, the IAU established a new working group on NEOs, which was supposed to act as an 'international focal point' for scientific knowledge on this topic.

There still was some rivalry between those who considered the danger of an NEO impact a potential end to the existence of humanity, and those who brushed it off as a hoax and PR stunt. The IAU did not have an official position on this issue. In 1998 it proved impossible to stay out of it, however. In March of that year, Brian Marsden issued a 'Minor Planet Circular', which implied that asteroid 1997-XF11 might collide with the Earth in 2028. Indeed, he explicitly did not rule out the possibility of a collision. Within days, earlier observations were identified that eliminated any risk, but by then the media had already exploded. Scientific explanations could not restore calm. As with the Pluto-as-a-minor-planet case, which happened around the same time, NASA and the IAU were caught completely by surprise. Neither the IAU General Secretary of the time, Johannes Andersen, nor NASA, which were suddenly flooded with phone calls from journalists, were amused. To prevent a repetition, the IAU announced that its working group on NEOs would 'in the future offer public peer review of all discoveries claimed to present a significant risk'.[22]

Another result of this incident was, however, that Andersen insisted on having a clear agreement about the status of the MPC. So far the relations had been founded on informal (i.e. verbal) agreements, but the media scare proved that the MPC had a public profile to match the rest of the IAU put together. After lengthy negotiations, proper *Terms of Reference* for the MPC were formally approved in 2002. A sticky bone of contention was the MPC data base of about 2 million 'single-night' observations of asteroids, which could be

---

[22] IB 86 (2000) 3.

connected with preliminary orbits of newly discovered objects and thus consolidate their orbits, but to which only the MPC staff had access (partly also because many observations were of military origin, made in the course of looking for satellites). But a database that has the IAU name on it must be accessible to the entire scientific community, and after extended and difficult negotiations, this was the final result.

The media frenzy surrounding XF 1998 put NEOs high on the political agenda, possibly reinforced by the release of two major Hollywood blockbusters, *Deep Impact* and *Armageddon*, on putative asteroid impacts, just months later. But as a result, the US Congress directed NASA to search and find 90% the NEOs larger than 1 km in size—and actually approved the corresponding funding. NASA established a Near-Earth Object Program Office at JPL in Pasadena, which cooperated closely with the MPC. This search is on-going and well advanced—at least in the northern hemisphere—with 90% of the NEOs as large as this already discovered, catalogued, and certified to be harmless for the next 100 years or so. The search is being extended to smaller NEOs. The southern hemisphere has as yet been more sparsely covered, but the southern asteroid detection capacity will soon be increased by the 8.4-metre southern Large Synoptic Survey Telescope (LSST, planned to be operational from 2023), which will survey 3/4 of the sky twice per week with its giant CCD camera.

The accuracy of the observations and orbit computations is illustrated by the fact that a small impact has already been followed with a sufficiently precisely predicted position for recovery of surviving pieces of the meteorite at sites in Sudan in 2008 and in Botswana in 2018. The scientific and technical difficulty in reaching 100% completeness especially concerns asteroids coming from the inner Solar System; they will approach the Earth in daytime, when they cannot be observed.

In the meantime, the IAU also became more active in the UN *Committee on the Peaceful Uses of Outer Space* (UN-COPUOS) itself, the main international forum to discuss the political and legal aspects space-related issues. As mentioned, the IAU had obtained official observer status in 1995. This position was primarily intended to call attention to the increasing problems of radio interference with astronomical observations, caused by communications satellites in space. But the IAU was not there to protect only its own interests; it could also offer assistance with other issues, including Near Earth Objects, a topic in which the diplomats were genuinely interested. To deal with the issue, a dedicated expert group called Action Team 14 (AT-14) was established, which included IAU representation and met regularly at the UN facilities in Vienna.

IAU President Catherine Cesarsky reading out an official statement on behalf of the IAU at the United Nations Committee for the Peaceful Uses of Outer Space. Photo: C. Madsen

Considering the multitude of questions raised by the NEO hazard—organisational, technical, scientific and, not least, legal and political—it comes as no surprise that it took quite some time for the initiative to work its way through the UN system, but the result was a set of recommendations to all UN member states on how to deal with this issue. These recommendations were endorsed by the UN General Assembly in 2013. The outcome includes the creation of an International Asteroid Warning Network (IAWN, colloquially known as 'I-warn'), acting as a clearing house for information on dangerous asteroids and any future terrestrial impact events and a Space Missions Planning Advisory Group (SMPAG, colloquially known as 'Same-page') that will coordinate joint technology studies for deflection missions and provide oversight of actual missions.

Hans Rickman, Quiet But Steady Progress

President of Commission 20 Positions & Motions of Minor Planets, Comets & Satellites (1997–2000), Assistant General Secretary (1997–2000), General Secretary (2000–2003)

Interview conducted in Vienna on 30 August 2018

Given the overall conditions in which the IAU functions it is hardly surprising that individuals can exert a strong influence its operations. Some Presidents and General Secretaries have been dynamic with marked opinions, others have taken a more quiet, but by no means less effective approach to running the Union's machinery.

Hans Rickman at the IAU General Assembly in Vienna. Photo: C. Madsen

Hans Rickman was born in Stockholm in 1949.

'I got interested in astronomy as a little boy of 11–12 years, when reading Donald Duck who went to Betelgeuse with the kids. Everything there was of course gigantic, and I found this very fascinating. But I never became an amateur astronomer.

At Stockholm University I studied mathematics and thought of becoming a mathematician. But I remembered my old interest in astronomy and took the first basic courses for fun. I was asked by astronomer Lars-Olof Lodén, if I'd like to help out with classifying objective prism spectra of stars in the Southern Milky Way that he and his wife Kerstin had taken in South Africa for a catalogue project. I said yes and found myself spending the summer days at a microscope in the Saltsjöbaden Observatory.

By that time, I decided to skip mathematics and go on with astronomy. I had heard about all the interesting things an astronomer would do, like going to foreign places to observe. For my PhD project, however, Lars-Olof suggested that I should do something different, namely, studying the Solar System and its origin. He sent me to Hannes Alfvén's lectures at the Stockholm Polytechnic and this was a great experience for me. Alfvén and his colleagues gave me the topic for

(Continued)

my thesis: The capture of comets, meaning the transfer of comets from long-period to short-period orbits by Jupiter's gravity.

So, that's what I stayed with during my whole career—I became a cometary astronomer and probably regarded the study of more remote and energetic objects with a similar reservation as the extragalactic astronomers looked at what I and my colleagues were doing. Moving to Uppsala, I met a colleague, Claes-Ingvar Lagerkvist, and we began to build up a small Solar System research group. In 1983, we decided to organise a meeting on asteroids, comets and meteors. For the latter, Bertil Anders Lindblad at Lund Observatory came on-board. Things had begun to happen around Comet 1P/Halley and ESAs Giotto mission, and so we invited people that we knew about and some 60 or 70 people came to Uppsala for this first asteroids and comets meeting. Then the Americans got interested and this way the seemingly never-ending series of international ACM meetings started.'

Rickman became a professor at Uppsala in 2000 and later also received an assignment at the University of Warsaw.

'I attended the Grenoble GA in 1976. I was still a student at the time and I found it really exciting. I particularly remember a talk by Carl Sagan. Then at the Montreal GA, I became a member of the IAU although I couldn't go to the GA myself. But I attended the one in Patras, in Baltimore, Buenos Aires, The Hague and so on. At the meeting in The Hague, I realised that Johannes Andersen was becoming the new Assistant General Secretary, but I had no idea that I would become one myself and thus succeed him. However, as it happened, for the Kyoto GA in 1997 the nominating committee proposed me and thus, I served as AGS until 2000 and then as GS between 2000 and 2003.

I felt that this was an honour that was being bestowed on me and that I should not say no to it. Indeed, I felt that if I could do something good for the IAU I was ready to do it. My university supported my decision, so that I was relieved of teaching duties during the three years of my term as GS.

A relaxed moment during the EC 71 Meeting in Paris in 1998: (left to right) Birgitta Nordström, Hans Rickman and Catherine Cesarsky, IAU Vice President at the time. Photo: Collection Hans Rickman

(Continued)

The task as Assistant General Secretary entailed quite a bit of work, I had to communicate with e-mails and letters with proposers of symposia, collate scientific reports and also dealing with grant recipients. Then there was the 1997 XF11 affair. Johannes Andersen handled it extremely well but given my own scientific background and expertise, I believe that I was of help to him in this matter.

During my term as General Secretary, I concentrated mainly on the follow-up and practical implementation of initiatives taken by my predecessors, especially those by Johannes Andersen. In this sense there was quite a bit overlap between us. For example, at the time, there was quite a fight between Brian Marsden, head of the Minor Planet Center (MPC) and parts of the scientific community about the access to its database. Since I was a friend of Marsden, I tended to defend him in the beginning, which wasn't popular in some corners. Marsden saw the MPC as the clearinghouse for observations and he felt that he had the responsibility to oversee the process and to screen and verify the quality of the observations reported. In other words, he would decide which observations the astronomers would be allowed to see and which not. Furthermore, some MPC datasets of military origin were classified. But despite my understanding for Marsden's motives, the demand for open access to a database in an organisation under the auspices of the IAU was of course non-negotiable.

The discussion had been on-going for quite some time and progress had indeed been achieved. But the final agreement had not been signed. In the end, however, we succeeded in our goal that all observations should be made publicly accessible and the contract was signed in 2001. This also paved the way for solving the question whether the IAU should continue its relationship with the MPC. In fact, until the Sydney GA, it was not clear that the IAU would continue to sponsor the MPC, but luckily, things were sorted out in the end.

As GS, Franco Pacini was "my" President, who, of course, will be remembered as the father of the International Year of Astronomy, which he formally proposed at the GA in Sydney. We became real friends and I have many good memories of him. Some Presidents are very much "hands-on", others see the task as one of "presiding" rather than of "doing". I didn't see Franco as a very active President at the time, which left me with a lot of freedom as General Secretary.

I typically spent two weeks at the time in Paris and all in all probably a third of my time there. I really loved the city, not least Le Village de la Butte aux Cailles area not too far from the IAU office.

At the beginning of my term, the Secretariat moved from its previous location on the 3rd floor of the Institut d'Astrophysique into new offices on the 2nd floor. This meant a gain of space in general—and one additional office. To my regrets, however, we lost the beautiful view towards the dome of the Paris Observatory, but you can't have everything.'

With the move, the office had finally gained sufficient space, but it didn't change the fact that a staff contingent of only two full-time people was small and—according to some—much too small. Since funding for additional staff was not available, a way to mitigate this apparently permanent problem has been to seek solutions in technology.

(Continued)

'A major concern was the membership database, partly because of the steady growth of the membership, partly because of the rapid progress in IT technology and capability, which meant a constant "catching-up" game. Add to this the fact that the database rested with the General Secretary and therefore had to be migrated with the change of GS—in fact like the physical office had moved in the past—and it was clear that we had to do something. We moved the database from Copenhagen and with the help of two IT support people from Uppsala, we set about to update and improve it. With this, the last edition of the Information Bulletin that I prepared was also the last one to be sent out in print version to everybody by regular mail. Instead, the Bulletin was put on the web for download. After my term, the IT support including the database was taken over first by a local company and subsequently by ESO, where it has found a permanent home.'

In the wake of the Balkans wars, in 1992, the United Nations Security Council had imposed a wide-ranging embargo on Yugoslavia and its successor state Serbia-Montenegro. The Security Council resolution, which is legally binding for all UN member states, also called for '[suspension of] scientific, technical and cultural exchanges and visits' and therefore Yugoslavia had been expelled from the IAU. However, by 2000 the situation had eased and Serbia-Montenegro was allowed to rejoin the UN.

'Like other scientists, I had been troubled by the boycott on scientific cooperation. Luckily the end to it came around the same time as I assumed the office of General Secretary and thus, it was possible to take action to help Serbia and Montenegro back into the Union, which took effect at the Sydney GA.'

Other challenges of those days included the fight against light pollution and radio interference, topics that are still on the agenda today. Furthermore, the question of the potential threat from Near-Earth Objects gained prominence—something that was close to the heart of Rickman. All of these topics, however, required close interaction with other international bodies such as the UN, ITU and ICSU.

'The IAU obviously has an important role to play here. Even so, as GS I was happy to see the IAU represented in those bodies by others than myself. For example, Johannes Andersen did a great job at the UN Committee for the Peaceful Uses of Outer Space, having gained experience at the forum from the 1999 UNISPACE III Conference.

Our relations with ICSU, however, once again came under question. Since the time that ICSU succeeded the International Research Council, IAU had benefitted from its financial support, but the subsidies had diminished. In fact, for 2002 and 2003, the IAU received no subsidies at all. Luckily, it was decided to maintain the

(Continued)

relationship and in 2004, i.e. one year after my term as GS had ended, the IAU became a leader of an ICSU project entitled "Comet/Asteroid Impacts and Human Society" in collaboration with COSPAR and other bodies. I had personally worked hard for this and so it was gratifying to see it being realised. Needless to say, given the interdisciplinary nature of the question how society should tackle this situation, not the least in a potential impact scenario, I believe that it was important to engage with and support the emerging efforts in that area.'

Clearly, the IAU can also play an important role in encouraging the development of astronomy in developing countries. It does so with great enthusiasm and vigour today, but also when Rickman was General Secretary.

'I had the pleasure of participating in the inauguration of a 45-cm telescope (under the TAD label) at the PAGASA Astronomical Observatory of the University of the Philippines in Quezon City. This was the largest telescope in the Philippines and as it was a donation by Japan, former IAU President Yoshihide Kozai was there, too. After the ceremony, I spent some time with the young people there— all astronomy enthusiasts. One of them asked me: 'Do you know that a comet was discovered in the Philippines?' And I replied: "Yes, Comet Boethin in 1975!" They must have thought that I as IAU GS would have known everything, but actually it was because I had published a paper on that comet! Be that as it may: It was an enjoyable experience and a pat on the shoulder to a country that had just joined the IAU.

After my term, it took time for me to come back to research, but the break also proved to be a source of inspiration. In fact, I believe that I've become a better scientist since those days.

I retired from Uppsala in 2014, at 65, then from Warsaw 2015 and then I thought that would be it. I tidied up the last loose ends, wrote a book about comets and, in general, prepared for a quiet retirement. But then the ARIEL mission was selected by ESA and my old institute in Warsaw asked me to come back to lead a research team on exoplanets research. You can say that towards the end of my professional career, I finally moved out of the Solar System.'

So despite his quiet ways and extremely modest personality, no quiet retirement for Hans Rickman. Not a bad thing for astronomy.

Oddbjørn Engvold: Network and Training for Young Astronomers, Impact on Human Culture

President of Commission 10 Solar Activity (1994–1997), President of Division II Sun & Heliosphere (1994–1997), Assistant General Secretary (2000–2003), General Secretary (2003–2006)

Interview conducted in Oslo on 19 June 2017

'"My" first General Assembly was in 1967 in Prague. Four decades later, my term as General Secretary came to an end at the 2006 General Assembly, again in Prague.'

Oddbjørn Engvold in front of the Svein Rosseland House at the Institute for Theoretical Astrophysics in Oslo. Photo: C. Madsen

'One thing that caught my attention at that first IAU General Assembly was that it provides astronomers with a unique opportunity to learn about and interact with other sub-disciplines in astronomy. I come from Division II and largely knew the 11 other divisions by attending General Assemblies, but only when becoming General Secretary I became fully aware of their significance. I knew about the various fields, but in 1967, the time of "my" first General Assembly, there were fewer opportunities to establish personal contacts with scientists from fields outside my own. And so, for young people and students, IAU General Assemblies offer unique possibilities to attend excellent presentations over a wide field of astronomy, which one does not get at topical meetings.'

'As IAU General Secretary one gets immediately involved in the publication of Proceedings from General Assemblies, the Highlights of Astronomy presented at the General Assemblies, as well as Proceedings from six symposia every year and its news bulletins. My involvement and responsibility for the Proceedings from the General Assembly in Sydney also became a quick and most helpful introduction to the many aspects of running an international union in science.

(Continued)

The two thick books from the XXVth General Assembly were the last in a series that were published by the Astronomical Society of the Pacific. The IAU Executive Committee decided to open up for a competition between several Publishing companies, which was finally won by Cambridge University Press (CUP). This transition involved a lot of meetings and discussions between CUP and the IAU. I had earlier gained some experience with electronic publishing from working with Kluwer publishing and I realised that "this was the future", although it had many implications. At the IAU, one of the first things we did in that direction was switching to web publishing of the IAU Bulletin. Also, the first IAU e-Newsletter—Volume 2006 no. 1—was sent electronically to individual members in April 2006.

Linked to this was the necessity to update the IAU web server and I also engaged a local company to modernise the web site and make it a key communication instrument for the Union. I found this challenging. The company did too! In fact, the company failed. Therefore, I had to involve a programmer, Laurent Pezeron, who fixed it to the extent that we had a functioning system in place for the Prague General Assembly. Still, this was an interim solution and soon thereafter it was replaced by a more satisfactory solution implemented by Lars Lindberg Christensen and his ESO-team.

The IAU is really run on a shoestring budget. The General Secretary and the officers are not paid by the IAU, so they rely on income from their home institutes. This means that the bulk of the IAU income goes back to the member countries. I believe this is a precious and politically prudent policy. It also means that the people who are really devoted and motivated will work for the IAU; interestingly, there is no problem to get people to work for the IAU.

I myself had full freedom to spend 100% of my working time on the tasks of the General Secretary. As Assistant General Secretary I had, of course, the chance to familiarise myself with the work, but still, when I came to the office in Paris, after the 2003 General Assembly in Sydney, it really started. There was much to learn. Quite frankly, it was a bit frightening and at one point I wondered why I had agreed to do it in the first place. This also means that the General Secretary is dependent on the help of the permanent or semi-permanent office staff. I noticed fairly early some frictions between the Executive Assistant and the younger more temporary staff. At the time, I put it down to trivial personality issues, but it certainly complicated matters. Seen from today's perspective, however, there might have been more to it than that. But other administrative issues caught my attention.

I was surprised by the fact that at the time we had three banks and I spent quite some time in sorting out bank matters. A Dutch bank, which IAU had used for some 70 years, was one of those that caused concern and problems. For historical reasons, we had several accounts at that bank. Having transferred funds from the wrong IAU account in the bank to cover expenses for the IAU General Assembly in Sydney, the bank demanded penalty charges, which by end of 2005 had become around 24,000 €. In the autumn of 2005, I was informed about the error of the bank that led to this mess. With local help from Eugene de Geus, whom I knew from working earlier with Kluwer Academic Publisher, and an English lawyer, Stephen Kahn, and personal meeting with representatives of the bank in Utrecht, the bank turned around and returned the "penalty charges" to the IAU account. But subsequently, all business with that bank was discontinued. Serious breaches at the local French bank that we used, which were discovered

(Continued)

in the first year of Karel van der Hucht's term as General Secretary, were similarly shocking but solved in the end.

In retrospect, in view of the later the "Orine affair", I've wondered if what I saw as a messy situation was really just the top of an iceberg, but at the time, I had no suspicion that there was more than a simple need for re-organising our accounts. Alas, there was. So, like other General Secretaries of that period, I've asked myself if I overlooked signs of irregularities or troubles. But I also see a system failure and a cultural issue.'

Monique Léger-Orine, Executive Assistant of the IAU Office, was the senior of the two only full-time employees of the IAU and therefore placed in a central position. In the General Assembly newspaper from Prague, IAU President Ron Ekers expressed it this way: 'Our secretarial staff in the Paris Office, ... who maintains the IAU corporate memory.'

'We IAU General Secretaries all come from science, which in many ways is a trust-based system. It can be so, because in science if you cheat and thus breach the trust, you're finished. Relating to the "Orine Affair", where you had to interact with people outside our own culture and working in a different set of functions, it was a severe weakness of the IAU. But I'm sure that we've learned the lesson. Otherwise I admire the way that the IAU has been dealing its many responsibilities—all the programmes that it has done—and it has been remarkably rewarding to have the opportunity to be involved and to learn about this, to see how it functions and how people go in there and do their very best whenever they have the opportunity.' Overseeing the internal administration is an important task for the General Secretary. Another one is maintaining relations with other international bodies.

'The fact that astronomy is connected scientifically to a number of other natural sciences and technologies makes it essential and of fundamental importance to maintaining established interactions with other unions and organisations in science. This was and is among the central duties that a General Secretary has to be acquainted with and to deal with timely and correctly. The foremost among these unions is The International Council for Science (ICSU, after its former name, the International Council of Scientific Unions), the umbrella organisation for international scientific unions and for national scientific bodies. I represented the IAU at a meeting of the ICSU unions in Suzhou, China in 2005, where "Natural and Human-Induced Hazards" became a new, high priority area among associated space science related unions and organisations. Also, in the wake of the attack on the World Trade Center on 11 September 2001 scientists from China and Arabic countries, in particular, experienced problems in obtaining visas in time to attend scientific meetings. ICSU's strong commitment to safeguard the freedom and responsibility in the conduct of science also became activated then.

We could also lean on ICSU for advice on other large-scale issues such as for instance the "two-China" issue. One time, we had failed in a couple of forms to present China and Taiwan as one country under the formula "China Nanjing" and "China Taipei". We received serious complaints, but we succeeded to resolve the controversy in a mutually acceptable way. We had to balance on a sharp edge. Science has a unique position in terms of interaction across political systems, but even so, it is not always straightforward, especially when it comes to totalitarian regimes, and it is here that organisations of high international standing, like ICSU, with its widely respected rules and principles, can play an important role.'

(Continued)

IAU's collaboration with 11 other scientific organisations like COSPAR, CODATA, SCOSTEP, etc. was equally important and essential for fulfilling its responsibilities to its members.

In the period 2003–2006, the 'status' of Pluto returned. Historically, the IAU has enjoyed the role of arbiter and custodian of astronomical nomenclature, including, of course, the bodies in the Solar System. Improved observational results during the twentieth century, including the discovery of 'new' objects beyond the orbit of Neptune, however, began to raise serious questions as regards the number of planets in the Solar System which in turn sparked a debate on what constitutes a planet. The consequence of this was either that the 'ninth planet', Pluto, would no longer be considered a planet or that the number of planets would increase significantly. A rapidly growing number of extra-solar planets was also being discovered, so timely action was clearly needed.

To this end, the IAU Division III established a working group tasked with providing recommendations to the Executive Committee for a resolution to be brought before the IAU General Assembly in Prague.

'But the working group could not reach an agreement. In fact, the group came up with three different definitions. Also, realising that we actually needed to broaden the participation of people from outside of astronomy, because it was likely to have a societal impact, we set up a special committee, with Owen Gingerich as chairman and participation by Richard Binzel, André Brahic, Dava Sobel, Junichi Watanabe, Ivan Williams and IAU President-Elect Catherine Cesarsky.

Richard Binzel of the Planet Definition Committee presenting the proposed 'Pluto resolution' to the GA participants. Photo: IAU Archives/L. Holm Nielsen

(Continued)

The outcome of their deliberations was eventually presented to the IAU members at the IAU General Assembly in Prague. At Prague, the discussion was chaired by Richard Binzel and Jocelyn Bell-Burnell. I must say, especially Jocelyn put up one of the most excellent performances that I've experienced, holding in her hand the iconic cartoon dog that carried the same name as the celestial body. It was fantastic. What followed was a very lively discussion—to put it diplomatically. In fact, there were strong reactions. As it appeared, many members were unhappy and expressed this in a very direct way. I was shocked because we thought that we'd handled it in the best manner, and so I and, I believe, the Executive Committee were taken by surprise that it would cause such a big discussion. Actually, the original proposal by the Executive Committee was not accepted. It was evident that there were lots of feelings involved, so it was not one of the usual purely rational discussions that are the norm within our discipline. In any event, a modified resolution was adopted by the General Assembly, putting a formal end to an unusually strong controversy.'

With the vote, Pluto was demoted to the less glorious ranks of 'dwarf planets', a decision that continued to meet resistance and protests, not the least in the USA, where it was even presented as an 'anti-American act'.

'Unsurprisingly, it also made lots of waves outside of the conference centre in Prague—in fact in the worldwide public. Luckily, I had appointed Lars Lindberg Christensen as our press officer. He could handle the press and that made my life easier during this critical discussion.

In earlier years, the IAU Office followed the home site of the General Secretaries, which gradually must have become quite challenging, related to a growing archive and the need for regular availability of administrative office assistance. The decision to move the General Secretary rather than her/his office, became inevitable and highly welcome. In my opinion, to have the IAU Office permanently located at the premises of Institut d'Astrophysique de Paris, at a vital and stimulating astronomical milieu, was a fortunate and lucky one. The only concern was that the General Secretary and the Assistant General Secretary, who need frequent and close access to IAU's Office, were usually located in European countries.

During my term, we also finalised a new contract with the Minor Planet Center. This had become necessary because the equipment at the MPC had become insufficient to handle the increasing amount of data that were coming in. Furthermore, there was an increased interest, perhaps even concern—as regards the Near-Earth Object issue. That included the media, and so the IAU also decided to appoint a small group of specialists to handle any public communication issues that might arise, with participation by David Morrison, Richard Binzel, Andrea Carusi and Don Yeomans. The MPC agreement had been underway for quite a while, but it was signed during my tenure.'

The concerns relating to the potential NEO-threats, including the public communication issue, has remained and been the subject of discussions at the level of the United Nations, with the participation of the IAU. Thus in 2013, the UN Committee on the Peaceful Uses of Outer Space adopted a set of recommendations to the UN member states, including the proposal to establish an International Asteroid Warning Network and Space Mission Advisory Planning Group among the space-faring nations.

(Continued)

'Perhaps the IAU activity that has been closest to my heart and most appealing has been the educational and outreach programmes, especially the International Schools for Young Astronomers (ISYAs). During my time as General Secretary, I attended an ISYA in Morocco and got an impression of what it meant to the young students. The 29 young people were so enthused—it was a gift to them and they were so excited, so devoted. However, despite the momentary excitement, the real challenge was to make this effort sustainable and achieve a long-term impact. That meant institutional, perhaps even government involvement. After my term as General Secretary, the Norwegian Academy of Science and Letters (NASL) began to provide some extra funding. The schools are clearly contributing really meaningfully to the overall educational fabric.

After my term as General Secretary I told the Norwegian Academy of Science and Letters about IAU's ISYA program and they took an interest in this and started to provide some extra funding.'

Thanks to the additional support provided by the NASL, it is now becoming possible to organise three ISYAs within a two-year cycle, with two funded by the NASL and one by the IAU. Recent ISYAs have been held in Iran and Ethiopia and the IAU has applications from several countries across the world.

ISYA participants in Egypt 2018. Photo: IAU Archives

'The schools are clearly contributing to the overall educational fabric in a way which is really meaningful.'

Over the last couple of decades, the IAU has begun to work with various philanthropic organisations interested in promoting science and technology. One of these is the Kavli Foundation, established by Fred Kavli, an American industrialist of Norwegian origin. The Kavli Foundation works with the Norwegian Academy of Science and Letters on behalf of which Engvold was asked to chair the committee in charge of the Kavli Prize for Astrophysics.

(Continued)

The links between the IAU, the NASL and the Kavli Foundation have been further strengthened, e.g. as the winners of the Kavli Prize may also be invited to give lectures at the ISYAs whenever it fits in the planned and agreed curricula.

'When you go and give lectures at the ISYA, you're not simply giving the students a gift. Talents in science exist everywhere, and by participating you become part of this worldwide network. I find the educational programs of the IAU to be among its most meaningful activities.'

Oddbjørn Engvold was born in 1938 in Askim in the southern part of Norway. The difficult times during the 2nd World War forced his parents in 1941 to move to the middle-east country of Norway, where he was raised.

'When I finished secondary school, I wasn't quite sure what to do. I considered seriously studying either Architecture or Medicine, but in the end, I did a bachelor's degree in Physics and thus became introduced to Solar Physics and Astronomy. When I came to the Institute, Svein Rosseland was still director of the Institute and he fascinated me greatly. During my study for the master's degree, I was given a part-time job at the then newly established Solar Observatory north of Oslo. I was thereafter offered a research assistant position. A little later, I was lucky to spend one year at Kitt Peak in Arizona and that opened up the entire field for me. A few years later I got a US scholarship to work at Sacramento Peak Observatory in New Mexico, which also became a most personally rewarding experience and the start of life-long fascinating collaborations with colleagues in USA, Canada, Europe and Asia, also involving frequent use of solar and stellar instrument both in space and ground-based.'

With this background, it is hardly surprising that Engvold's scientific interests have focussed on Solar Physics and, among other tasks, he was involved in the early design phases for the Large European Solar Telescope (LEST) project. He became a professor at the University of Oslo in 1989. Now an emeritus professor, he enjoys the opportunity to concentrate professionally on his science. But he has not lost a sense of the bigger perspective:

'Astronomy has impacted human society in many direct ways. It is interesting to note that when the Union between Norway and Denmark was dissolved in 1811 and Norway had aspirations to become a sovereign nation, one of the first actions was to establish a university in Oslo. And the first building of the new university was devoted to astronomy. In those days, astronomy was important for a sea-faring nation like Norway, but other practical needs had to be catered to as well in terms of geographical mapping, almanacs, etc. Speaking more broadly, astronomy is rather unique in the way that it connects many with scientific disciplines, it's more accessible to people than many other sciences and it has influenced human culture—our understanding of the physical world, but also religion and philosophy—more than perhaps any other human endeavour. Thinking of the great paradigm shifts in astronomy regarding the importance and position of the Earth, I believe that it has had a sobering effect on humanity. To me, astronomy has had a much stronger and more positive impact on human society than we normally realise. It lets us appreciate the richness of the Universe while also fostering rational thinking, which I believe is a need for humanity to survive. And I believe that the IAU has played its role in this context as the central global networking setup for professional astronomers and also as an interface with society at large.'

Karel van der Hucht: A Tougher Job than Expected

Assistant General Secretary (2003–2006), General Secretary (2006–2009)
   Interview conducted in Leiden on 19 April 2017
   General Secretaries of the IAU have to cope with a heavy workload. Karel van der Hucht knew this in advance of his term (2003–2006 as Assistant General Secretary, 2006–2009 as General Secretary), but still, the job proved much tougher than expected.

Karel van der Hucht during a visit to the Leiden Observatory. Photo: C. Madsen

'In 1972, I completed my studies with Kees de Jager, and he offered me a job at the Space Research Laboratory in Utrecht. My own scientific interest was Wolf-Rayet stars, thanks to Peter Conti, who for some time came to Utrecht to teach Wolf-Rayet stars. At the Space Research Laboratory, I participated in space research projects with balloons and satellites etc. One of the objects for which we obtained a UV spectrum was γ Velorum, the brightest Wolf-Rayet star. I already had an invitation from Peter Conti to come as a fellow to the Joint Institute for Laboratory Astrophysics in Boulder, Colorado, upon completing my PhD, and with that spectrum, I made my "entrée" there in 1978. It was during that time, working with Peter, that I had my first encounter with IAU matters.

Together with Bert de Loore, Peter Conti was involved in the organisation of IAU symposia on hot massive stars. The first one that I helped to organise was IAU Symposium No. 143 in 1990. It took place in Bali with Bambang Hidayat in charge of the local organisation. I also became involved in the proceedings and I quickly understood the importance of having the proceedings published as soon as possible. With Bambang, we got the publication out in half a year. Later IAU Symposia I was involved in took place on Elba (163), in Puerto Vallarta, México, (193) and on Lanzarote (212).

In 2001 I got the telephone call from Franco Pacini, who was President of the IAU at the time, asking me if I would be willing to serve as IAU General Secretary. Maybe he was desperate—I don't know how many he'd asked—but I said "yes"!

(Continued)

I soon met with the IAU officers at the time, Hans Rickman and Oddbjørn Engvold, who would eventually become my predecessors. And then in 2003, I began as Assistant General Secretary. I would be in charge of guiding the organisation of the IAU colloquia and symposia, a task with which I was of course quite familiar.

At that time, there had been a change of the IAU publishing partner—from Kluwer to the ASP, the Astronomical Society of the Pacific. They did a good job, but I soon found out that for a couple of years, many proceedings had not yet been finalised. So I got a list of these proceedings, some going back to the time of the General Assembly in Manchester in the year 2000. I had to chase down authors and editors. Most of the work of an editor is, in fact, chasing down authors, putting pressure on them, setting deadlines and so on, and eventually, I managed to overcome the deadlock. In the meantime, Oddbjørn had begun to argue in favour of electronic publishing and the ASP was not yet ready to go along with this. Cambridge University Press, however, was willing and so, once more, the IAU changed its publisher. Oddbjørn also wanted to introduce a peer-review system for the IAU publications, but eventually, the idea was dropped. And then, at XXVIst General Assembly in Prague in 2006, I took over as General Secretary. Oddbjørn and I met shortly afterwards at the IAU Secretariat in Paris. The meeting was brief. Though Oddbjørn was always available for advice, I was now on my own, his parting words "You can always ask Monique Orine, the Executive Secretary," still ringing in my ears. And he added: "Everything is based on trust".'

One of Engvold's initiatives had been the introduction of the IAU website. Yet its implementation by an external company was not without problems. At a point in time, Lars Lindberg Christensen from ESO approached van der Hucht with suggestions for a much-improved website, to be prepared under a contract with ESO. The links with ESO in the widest sense were strong in those days, with Ian Corbett, who had recently retired from ESO, becoming Assistant General Secretary and Catherine Cesarsky, in her last year as ESO Director General, the IAU President.

'It was interesting to work with Ian Corbett and Catherine Cesarsky, who of course knew each other well from ESO. Now the three of us had to work together with two major tasks: The forthcoming XXVIIst General Assembly and the International Year of Astronomy (IYA 2009), which obviously was a huge challenge not the least in terms of budget.

In these circumstances, my initial ambition was not to make specific changes in the IAU, but to make things running as I thought they should run. That of course also included the timely publication of the Proceedings and Highlights of the Prague meeting, the IAU Information Bulletin and keeping feeding content into the website.'

(Continued)

When it comes to 'running the IAU', the burden is squarely placed on the General Secretary, but no salary comes with the job, so that the General Secretary will be dependent on other income, in most cases from a 'day-time job'. This also means that quite a bit of work—the day-to-day operations—will be handled by the Executive Assistant, who runs the IAU office in Paris.

'Monique Orine was always in the Paris office—the first to come into the office in the morning, the last one to leave. I could always come in and ask her questions. With 20 years of service to the IAU, she had become the stalwart of the IAU Secretariat.

Then, in December 2007, we had a Christmas dinner together with the staff of the Secretariat. By the end of the dinner, I had the feeling that she wanted to tell me something, but I had to leave. I never saw her again. In January, I spoke briefly on the telephone with her. She was at home, clearly quite ill. When I returned to Paris during the following week, she had passed away in a hospital. We were all shocked and saddened. And it obviously left us with a huge problem. Monique had been central to the work of the office and now she was no longer there. Even her computer was locked.

We were very lucky to quickly being able to hire Vivien Reuter as a replacement for Monique. Vivien had her work cut out. Indeed she had to start from scratch. But she soon realised that something was seriously wrong. As it turned out to my horror, over an extended period, Monique had embezzled IAU funds, taking advantage of the trust placed in her, but cleverly also evading discovery through the external audit and the bank controls.'

For someone as honest, thorough and meticulous as van der Hucht, the revelation of financial irregularities must have been a true nightmare.

'I went to the IAU President, Catherine Cesarsky, to offer my resignation. It was my responsibility. But she rejected my offer. I also had great support from the Chairman of the IAU Finance Committee, Paul Murdin.'

After all, and whatever the circumstances, the problem was discovered on his watch and van der Hucht, barely half-way through his term as General Secretary, could hardly have prevented it from happening, given that it had apparently lasted for a long time and been conducted with great cunning. The challenge, then, was one of recovery—to re-establish trust in the functioning of the office and its oversight and to recover the lost funds as far as possible. And it had to be brought into the open—at the General Assembly in Rio in August 2009.

The General Assembly in Rio was arguably one of the most important assemblies in the history of the IAU. On the Agenda: The International Year of Astronomy and the new IAU Strategic Plan. The IAU was on a roll. Yet for the General Secretary, the smooth running of the meeting was the primary task at hand and that included many practical concerns, including security.

(Continued)

Young Astronomers Lunch during IAU GA in Rio. Travel grants are essential for them to attend a General Assembly. Photo: IAU Archives

'The IAU provides grants to enable young astronomers to participate in the General Assembly. These grants had to be distributed during the meeting, and for that purpose, in Rio, we had to go to the Rio branch of our bank connection. We needed to dispense the grants in cash, partly to avoid heavy bank fees, partly because some of the young participants might not even have a proper bank account. Vivien and I went to the bank and picked up the money. Then we took a taxi to the conference centre. On the way, in one of the tunnels in Rio, we got caught up in a traffic jam, but after 20 min, traffic started to flow again. The following day we learned that criminals had blocked the exit to the tunnel and systematically robbed the first twenty or so cars in the queue. Luckily, we were not among them.'

But van der Hucht had additional things on his plate.

(Continued)

UNESCO Director General Koichi Matsuura and IAU General Secretary Karel van der Hucht signing the cooperation agreement between the two organisations. Photo: IAU Archives

'As a General Secretary, I wanted to strengthen our links with some of the other international bodies, such as ICSU and UNCOPUOS—the United Nations Committee on the Peaceful Uses of Outer Space—though I did not manage to do as much as I'd liked. But we did conclude a Memorandum of Understanding with UNESCO about astronomical projects under the World Heritage Convention.

In the end, it was decided that Ian Corbett should represent the IAU at ICSU while I went to UNCOPUOS. One of the subjects that UNCOPUOS was dealing with was the Near-Earth Object (NEO) hazard, for which a working group had been established with the task of elaborating recommendations to the UN member states to enable proper response measures in case of a threat to the Earth of an asteroid strike. That also included the call for more observations of these objects. It was a nice working group with good people. Though it was not really my field, I began to take considerable interest in the topic. In the period 2009–2012, when I served as an advisor to the IAU Executive Committee, I continued to represent the IAU in the working group and even afterwards for some time. I remember one dinner for the group, on 14 February 2013. We discussed how to raise the awareness of the topic, and Thomas Jones, a US astronaut who was a member of the working group, jokingly exclaimed: "We need an impact!" Then on the next morning—literally out of the blue—the Chelyabinsk event happened, when a 20-metre sized near-Earth asteroid entered the atmosphere above Russia and, despite its relatively small size, caused havoc as the shockwave enveloped a large swathe of land in Siberia causing both human injuries and material damage.

I started to compile a chronology of asteroid fly-bys between the Moon and the Earth from 1800 to 2200 and a list of relevant scientific papers and meetings, which became quite popular on the IAU website. In the meantime, it has grown enormously—it's now more than 600 pages—and in fact, I'm still keeping it up-to-date and on-line, now at the ESA website.'

Van der Hucht was born in Nuth, the Netherlands, in 1946, and embarked on his studies in physics in 1964, at Utrecht University. 'Some three years earlier, a

(Continued)

school friend had introduced me to observational astronomy by letting me observe the heavens through his hand-made telescope, one cold winter night. From that night on, I was totally dedicated to the idea of becoming an astronomer.' Retiring in 2001, he had spent 41 years at the Laboratory for Space Research (later named the Space Research Organization of the Netherlands (SRON), with annual teaching periods at the ITB Observatorium Bosscha in Lembang (Indonesia) between 1980 and 2000 as well as his long service to the IAU.

'The IAU has played an important role for me, and on the other hand, I've tried to do my best for our Union.'

Ian Corbett: Experienced Implementer

Assistant General Secretary (2006–2009), General Secretary (2009–2012)
Interview conducted in London on 6 July 2017
'Within a period of six years, the IAU changed quite dramatically. Perhaps this has become even clearer with time. We had restructured the way it handled its interfaces with both its own community and the outside world.'

Ian Corbett at the library of the Royal Astronomical Society in London. Photo: C. Madsen

Ian Corbett studied physics at Manchester and Oxford, obtaining a DPhil degree in 1965. He originally set out to become a theoretical physicist but gradually moved via experimental particle physics to technology development and science management. He stayed as a lecturer at Oxford until 1971, then went to CERN for two years, before returning to the Rutherford Lab. In 1980, he again went to CERN for two years, working on his own experiment and then the LEP project. On returning to Rutherford he became Head of the Applied Science Division. That exposed him to instrumentation projects in astronomy and space science, including the construction of the JCMT telescope in Hawai'i. Late in 1989, he was put in charge of the UK Science and Engineering Research Council's

(Continued)

astronomy programme and soon thereafter also particle physics. From there it was a short step to become Director of Programmes of the successor organisation, PPARC, in 1994. He retired from PPARC in 2001 and joined ESO initially as Head of Administration and subsequently as Deputy Director General.

'I had attended various IAU meetings because of my role in UK astronomy, including the General Assembly in Manchester in 2000. Years later, after I'd joined ESO, Catherine Cesarsky, then IAU President-Elect, encouraged me to attend the Prague General Assembly in 2006. Shortly afterwards, then President Ron Ekers called, asking me if I would be willing to serve as IAU General Secretary.

The new IAU core team at the Prague GA: Karel van der Hucht, Bob Williams, Catherine Cesarsky and Ian Corbett. Photo: IAU Archives/ESO/E. Janssen

At Prague, I was then elected as Assistant General Secretary for 2006–2009. Karel van der Hucht was the new General Secretary. We realised that we actually had complementary skills and we not only worked very well together, it was also fortuitous given the challenges that we were to encounter following the unexpected death in 2007 of the long-serving office administrator, and the subsequent discovery of her extensive fraud.

I had considerable administrative experience. The bulk of my time as Assistant General Secretary was spent in organising the Symposium programme of the IAU but I was also able to assist in preparing what we were going to do regarding "rebuilding the office". That meant completely changing "the system". Karel focussed on sorting out the immediate problems in the wake of Monique Orine's

(Continued)

death. Together with Vivien Reuter, appointed as Monique's successor soon after her death, he had made very considerable progress by the time I became General Secretary. We had closed the account with the bank that had been so careless in the handling of cheques and tied up other bank accounts and cash holdings. And, importantly, we installed a proprietary accounting software system which enforced transparency and ensured that the General Secretary could monitor all transactions.

Then we started two recovery actions—one against the bank, the other against the auditor, whom we felt had not performed in a way that we were entitled to expect. After tense negotiations with the bank, we arrived at a reasonable settlement. The action against the auditor, however, ran into the sand, but meanwhile, we had appointed another—and I believe—much better auditor. Vivien Reuter proved an excellent administrator and completely transformed the IAU office. And by the time that I finished as General Secretary in 2012, we had put in place a totally transparent system: We had a double-signature system for authorising payments, everything was online so there was a complete record of everything, and so on. In short, we had established a rigorous but transparent system. In addition, over time we moved the Union's reserves, which were approximately the equivalent of one year's dues, into various interest-bearing accounts.'

Although the issues with the office appeared to dominate, there were two other major activities in this period which changed the external face of the IAU. The period 2006–2009, when Ian Corbett served as Assistant General Secretary, was the time when the IAU, with growing intensity, prepared for and subsequently implemented the 2009 International Year of Astronomy (IYA), endorsed by UNESCO and the UN.

'The IYA was of course run by a big team, but it was useful to have people within the IAU Executive to follow this closely and that was primarily Catherine Cesarsky and then myself. My IYA involvement was basically behind—or inside—"the system", among other things in fundraising. But I should make it clear that the success of the IYA was down to many people worldwide and the inspirational leadership of Lars Christensen and Pedro Russo.'

Towards the end of IYA2009, i.e. at the General Assembly in Rio, George Miley presented the proposal for an IAU Strategic Plan. Its realisation would fall to the incoming Executive Committee and, of course, to Ian Corbett, the new General Secretary.

'I supported this initiative vigorously, especially as regards the setting up of the Office of Astronomy for Development (OAD). I was quite heavily involved in what you could call the administrative and development side of the OAD, including the selection of its location and the negotiations with the South Africans.

Looking back, for me, the big successes of the IAU during that period were the IYA, the Strategic Plan approval and the setting up of the OAD. There were other outreach-type activities going on within the IAU, some in fact since many years—the ISYAs and so on—so in order to coordinate all these efforts IAU President Bob Williams, George Miley and I set up a little oversight group for these activities. It was part of the concentration of these important activities and you could say that in this way, we ensured that the IAU had both responsibility and visibility whereas in the past there had been responsibility but rather less visibility.'

More changes were in the pipeline, not the least a major revision of the Division structure, which had been discussed by the Executive Committee several

(Continued)

times in the past as not truly reflecting the structure of twenty-first-century astronomy.

Ian Corbett speaking at the IAU General Assembly 2015 in Honolulu. At the centre is Norio Kaifu (IAU President 2012–2015) and on the left is Thierry Montmerle (IAU General Secretary 2012–2015). Photo: IAU Archives/R. Fienberg

'Thierry Montmerle, by then Assistant General Secretary, was keen to get his teeth into this. He did a lot of work and gave it real momentum so that by the time we got to the General Assembly in Beijing in 2012, we had the new Division Structure evolved after extensive discussion with the Divisions. This was subsequently followed up by a revision of the Commission structure. Thierry did a good job, consulted widely and with the active participation of the Commissions themselves. I think that it was a consensus opinion that what was bound to be a difficult exercise had been successful.

We'd also worked intensely with the national members. Some national members had increased their contributions, we'd dealt with national members that had serious problems honouring their commitments and we had new members, at least partly thanks to the IYA.

For the individual members, we instigated electronic voting and significantly increased the number of votes being cast and thus the involvement of the members.

Between Rio and Beijing, we'd also done some rethinking about the way the General Assembly was organised. We increased the number of invited discourses and the number of symposia and had a "general interest talk" from each symposium. The idea was that there should be "something for everyone" across the field of astronomy. If you go back to the initial idea of the IAU General Assembly, aside from the legal requirements governing decisions, it was to expose the community to the full but ever-increasing breadth of astronomy.'

(Continued)

Ian Corbett's term as General Secretary ended with the 2012 General Assembly in Beijing. Not just for the IAU, but also for the People's Republic, this was a prestigious event. IAU General Assemblies had frequently been graced by holders of the highest government offices and this would also be the case in China. At Beijing, much secrecy surrounded the run-up to the event.

'Ahead of the General Assembly, I made three trips to China before the General Assembly. The Chinese were very straightforward to deal with, but this was a big meeting for them, too—certainly in prestige if not necessarily in numbers. And, of course, their system is different from ours.

On Thursday before the General Assembly was due to start on the following Monday, I received a call from Gang Zhao in Beijing, asking about a re-arrangement of the inauguration programme, because "a very important person from the Chinese government would be in attendance". We obviously agreed to do this, but it involved not only the programme itself but also lots of protocol issues. Only on the Sunday were we told that "the very important person" would be Vice-President (and now President) Xi Jinping. In the end, it all went very smoothly. The Vice President was very charming and gave an excellent speech supporting international collaboration in science.

Opening ceremony, IAU GA Beijing, 2012. Photo: C. Madsen

We stayed on in Beijing for a few days after the General Assembly to wrap up things. It was an intensive time, but things went well.

In conclusion, by Beijing in 2012, we'd transformed the office, held the enormously successful IYA, established the OAD, changed the Divisional structure and improved communications with, and the involvement of, the ever-growing membership of the IAU. There was more to be done, but we'd made a good start.'

Lars Lindberg Christensen: A Public Window for the IAU

Chair of Commission 55 WG Public Outreach Information Management (2009–2012)

Interview conducted in Garching on 27 July 2017

In the wake of the IYA and with the establishment of the OAD and OAO, the IAU has increased its public visibility considerably. But already in 2006, the IAU engaged a dedicated press officer—Lars Lindberg Christensen. Says he: 'Of course, it was a tonne of work. Those were fun years'.

Lars Lindberg Christensen in his office at the ESO Headquarters. Photo: C. Madsen

'In late November 2005, Oddbjørn Engvold called me. He was keen that something be done because of the planet definition issue that would be a hot topic for the General Assembly in Prague in 2006. I was asked to prepare a proposal for the media aspects of the General Assembly and was subsequently appointed as a voluntary press officer. At Prague, we set up a press office, building on the experience mainly of Stephen Maran and Rick Fienberg from the AAS.

The media found out rather late what was happening, but at that point the interest just sky-rocketed. In fact, a few hundred media representatives arrived in Prague—far more than is normal for an IAU General Assembly—and there was a very substantial interest from the outside world in general.'

As the discussions evolved and draft resolutions changed almost by the hour, the press office was busy keeping the media informed. The internal conflict within the astronomical community could not be denied, so the only feasible strategy was to be as open and transparent as possible.

(Continued)

'We knew that the IAU would not necessarily be popular. However, the task of the Union is not to be popular, but to take useful initiatives on behalf of the scientific community. Despite the turbulent days, it was a great experience to be part of it. I had a hotel room on top of a hill outside of the city and every night, I could drive "home" to the peaceful hillside, look down on the town and reflect on the hustle and bustle of the day's Pluto events. Eventually, the final vote was successful, and the turbulence gradually died down.'

For Christensen, that was his first experience in the line of fire for the IAU, but not his first encounter with the IAU.

'Actually, my involvement with the IAU began in the late 1990s, when Johannes Andersen was serving as General Secretary. The tasks were to redesign the Information Bulletin and also the first IAU webpage, which we structured as a simple set of placeholders.'

At the time, Christensen worked at the Planetarium in Copenhagen, having graduated from the University of Copenhagen in 1996. In 1999, he joined the ESA/ESO European Coordination Facility for the Hubble Space Telescope (ST-ECF), hosted by ESO in Garching as an outreach officer. Here, Christensen and his team were asked to take over the IAU website. It had been done by an external French company, but it had become outdated, it was difficult to change the code and the documentation was poor, so, a complete overhaul had become necessary.

'This was a substantial reverse-engineering task for the team at the ESA-Hubble site at ESO. We put together the new webpage and especially the new database in a proper way. The database is a big thing because it's the substance of what the IAU really is, but had never been done in a systematic and modern way. It's the 12,000 plus current members as well as many "historical members"— with different capacities, different roles, different membership, changing address ever so often, and all of this has to be kept track of and presented in a good way.

It's important to realise that because the IAU is a volunteer organisation, the continuity and the business memory is incredibly difficult to preserve. The collective memory is not with individual people, but rests in the (digital) archive.'

The Prague General Assembly also marked the beginning of the serious preparatory work on the International Year of Astronomy, IYA.

(Continued)

Beginning with a conference at the US National Academy of Sciences in 2003, entitled 'Communicating Astronomy to the Public', an IAU Working Group on that subject was established. After a follow up conference at ESO in 2005, at the 2006 General Assembly, the Working Group was transformed to become Commission 55 (later C2). The photo here shows Pedro Russo at the CAP meeting in Athens in 2007. Since 2007, it has also produced its own journal, called the CAP journal. Photo: D. Crabtree

'It was at the same time that we started IAU Commission 55 "Communicating astronomy with the Public", and in that context, we could provide early help to the IYA preparations. Of course, the incoming IAU President, Catherine Cesarsky, was very keen to see the IYA happen in a proper way and upon application, a dedicated IYA secretariat was set up at ESO with a staff of two people, Pedro Russo and Mariana Barrosa. They became completely integrated into the out-reach team at ESO. It was a good way to do it, because they could then draw on the collective skills and experience of the entire group. Also, we could use an already existing architecture and structure for the IYA website—the same used for the IAU, ESO and ESA/Hubble.

Furthermore, in the wake of the Prague General Assembly, an IAU working group was also set up to coordinate the IYA preparations, chaired by Catherine Cesarsky. It was a working group that really worked! We also organised a meeting at ESO for the IYA national contact points in March 2007 to ensure that everybody had as much information as possible at an early stage.

In parallel, outside of the secretariat, the political lobbying at the UN level began with Catherine Cesarsky, Claus Madsen, Enikö Patkós, Norio Kaifu and others pulling the levers.

For the IYA secretariat, the work started for real in 2008, after the UN approval in late 2007. A major early task was fund-raising, partly for staff cost, partly to provide seed funding for the major activities, the so-called Cornerstone Projects. All in all, almost 700,000 € was raised.

(Continued)

In the Spring of 2007, the IYA 'SPoCs' gathered at ESO Garching for a briefing on the plans for 'The Year'. Photo: ESO Archive

It was a great and truly international undertaking with many interconnected activities. For example, one of Cornerstone Projects operated by ESO connected 84 observatories across the world in a continuous webcast. All our friends and colleagues—and sometimes competitors—came together. It was great to be part of this global activity, being together in this without politics or other frictions that occur in other human areas.

Also, books and movies were made and in general, the IYA offered a whole cornucopia of exciting activities.

In 2010, Pedro Russo oversaw the final activities of "The Year", writing the 1433-page Final Report and closing the accounts. But thankfully, the IYA created important legacy activities. Many projects have actually continued and there is still momentum to build on as the IAU sets out to celebrate its 100th anniversary.'

Christensen still serves as the IAU press officer, looking after the media, especially at IAU General Assemblies.

'Compared to many other scientific meetings, the General Assemblies are essentially review-type meetings. Few people pitch their latest discoveries and hence, the hot news are rare, although the IAU is improving in this regard, with (Nobel) prizes, etc. Therefore the media interest is normally more limited, but still, we typically have about 100 media representatives, including many dedicated science journalists, at a General Assembly and we try to help them as much as we can.'

(Continued)

'Brothers-in-arms': Lars Lindberg Christensen and Pedro Russo, here seen at the IYA Inauguration. Photo: IAU Archives

6

The IAU in the Twenty-First Century (2001–2019)

Turning Outwards

In 2012, IAU President Robert Williams said it well: 'The IAU is going through a period of transition, from an organization that historically has maintained a largely internal focus emphasizing meetings and events for its members, to one that is becoming more involved in education and outreach to the general public.'[1] Looking back seven years later, it appears that the Union has changed more in the past two decades than in any period before. It retained all its previous roles, but new activities draw most of the attention. The IAU has started to promote astronomy to a much wider audience. This was partly planned, partly it came as a reaction to outside pressures.

The highly public Pluto debates of 2006 and the hugely successful International Year of Astronomy in 2009 were major turning points. From now on, outreach, education and development, which had been started as small activities next to 'real' science, took centre stage. The Strategic Plan of 2009 formalised the new strategy. Since then, the IAU has opened three dedicated professional offices: Astronomy for Development (South Africa, 2011), Outreach (Japan 2012), and Young Astronomers (Norway, 2015). An office for Astronomy Education is being planned. With the recent campaigns to solicit public suggestions for naming exoplanets, one of the traditional core activities of the IAU (coordinating nomenclature) has been connected to outreach as well.

---

[1] GA newspaper 2012, 1.

© Springer Nature Switzerland AG 2019
J. Andersen et al., *The International Astronomical Union*,
https://doi.org/10.1007/978-3-319-96965-7_6

The IAU had supported development in non-member countries before, for example with the Teaching Astronomy for Development and ISYA programmes, but for the first time, it now has an active policy to stimulate its own growth. With now more than 13,500 members from 82 member countries, the IAU already includes most senior professional astronomers in the world. From 2018 onwards, it will also include junior members. The potential for growth, up to full IAU membership, is undoubtedly greatest in new countries, like in East Asia, but the development programmes are emphatically aimed at more than just astronomy. The underlying ideal, according to (Japanese) IAU President Norio Kaifu, is to promote astronomy as a way to stimulate a broader scientific, technological, economical, and cultural development of the whole host country.[2]

---

[2] IAU Strategic Plan 2010–2020.

The most important international astronomy meetings are undoubtedly the triennial General Assemblies. Since 2000, there have been General Assemblies in all continents except Africa. Busan, Korea, will host the meeting in 2021, with Cape Town, South Africa, to follow in 2024. The General Assembly in Beijing in 2012, in the enormous Olympic convention centre, marked a new increase in scale. It included eight Symposia, two more than usual. The meeting was opened by the then Vice President of China, Xi Jinping, and attracted more than three thousand participants. The opening of the 1997 General Assembly in Kyoto, Japan, had already been attended by the Imperial Couple, so the spectacular development of astronomy in East Asia has been appropriately recognised.

Xi Jinping, then Vice-President of the People's Republic of China, delivering his opening address at the 2012 GA in Beijing. Photo: C. Madsen

The past decade also witnessed a complete overhaul of the division and commission structure—after almost seventy years of often fruitless debates. In the meantime, the Symposium series—the 'Scientific flagship' of the IAU—is as active as ever; the 'Under One Sky' symposium at the General Assembly in Vienna in 2018, also dedicated to the centenary of the IAU, was number 349.

The new century began with the September 11 attacks in New York and Washington in 2001, which ended a decade of relative geopolitical peace (despite bloody regional conflicts in, among others, the former Yugoslavia, Rwanda, and Congo). The IAU reacted by proposing that three minor planets were named Compassion, Solidarity and Magnanimity. This well-meant gesture was mostly received positively, but inevitably also criticised: why now, and not after any number of other atrocities?[3]

As the 'fight against terror' continued, visa trouble was back on the top of the agenda of anyone who wanted to organise international meetings. Astronomical meetings that involved discussions of space technology were especially impacted, because of the sensitive nature of the technology—not just for security reasons, but also because of the intensifying technological and economic competition between, especially, the United States and China.

Local activists seized the opportunity of the GA in Honolulu to air a number of grievances and political views, including their resistance to the construction of the Thirty Meter Telescope on Mauna Kea. Photo: C. Madsen

The most visible political incident was probably during the General Assembly of 2015 in Hawai'i, where a local population organised protests against plans to build the Thirty-Meter Telescope on Mauna Kea, a mountain considered to be sacred by some, and also a natural preserve. The IAU decided

---

[3] IB 90 (2001); correspondence in IAU Archives 16C, file 'Working Group on for Planetary System Nomenclature'.

to avoid a direct involvement, treating it as a local conflict, in order to prevent further escalation. This seemed to work—there were no disturbing incidents during the meeting—but not everybody understood why the International Astronomical Union should not defend the interests of astronomy in this case. The eternal key question is, of course, how that is best done.

Aside from this, there were few divisive political issues. The IAU generally received political support from all sides, and it could welcome many new members. Even North Korea (re)joined in 2012 (it had formally been a member before, but in practice never sent delegates to a General Assembly). Of course, incidents always occur—astronomers get into political trouble in several places, every time putting the Executive Committee in a dilemma whether a formal statement would help or not.

On the scientific level, the first two decades of the twenty-first century saw the continuation of the developments that had started in the 1990s—somewhat similar to the way the 1970s and 1980s built on the discoveries of the 1960s. The telescope boom of the 1990s continued, for example, as several more 8–10 metre optical telescopes became operational. Around 2000, ESO opened the largest ground-based optical telescope system to date, fittingly but unimaginatively named the Very Large Telescope (VLT). It consists of four 8-metre optical telescopes, with several smaller supporting telescopes. In retrospect, however, the success of the VLT, including its project management, enabled ESO and its European member states to take the lead in subsequent international projects, most notably in the radio telescope project ALMA, designed to observe (sub)millimetre wavelengths, a part of the spectrum between radio and infrared that was relatively unexplored. This is a truly global project, built by a consortium of American, European and Asian institutions. Built in the extremely dry Atacama Desert in Chile, it became operational in 2013.

The ALMA Observatory at its 5000-metre high site on Llano Chajnantor in the Chilean highlands. Photo: C. Madsen

In the meantime, a series of large space observatories provided high-quality observations in various parts of the spectrum, from infrared to gamma ray. On the ground, new radio telescopes were built to observe the longest wavelengths, in Europe and the USA, but also in South Africa, India and China. This meant that the whole spectrum was opened for observations—a signal for astronomers to start thinking about a new generation of even larger instruments. The most eye-catching projects include several Extremely Large Telescopes of 30–40 metre size, the James Webb Space Telescope—the projected successor to Hubble Space Telescope—and the Square Kilometre Array (SKA) radio telescope, to be built partly in South Africa and partly in Australia by a global consortium. (A first Memorandum of Understanding, establishing an international steering committee for the SKA, was signed at the IAU General Assembly in Manchester in 2000). These projects are now in various stages of development; they are planned to start operating in the 2020s.

Besides the general observatories, there were many specialised space missions as well. Planetary missions were the most eye-catching, including a series of Mars rovers and the New Horizons mission to Pluto and the Kuiper Belt. Thanks to the Kepler exoplanet finding mission (launched 2009) and other search programmes, the number of known exoplanets is growing fast, approaching 4000 in 2018. Exoplanet studies have become a fast-growing field of astronomy, and a significant part of the scientific rationale for developing extremely large (optical) telescopes.

WMAP (launched 2001) and then Planck (2009) have produced ever more detailed maps of the Cosmic Microwave Background radiation, following the earlier COBE mission. The HIPPARCOS astrometry mission is now followed by Gaia (launched in 2013), which will provide high-precision data of more than one billion stars and other objects. Gaia's second data release in 2018 was already so impressive that overnight, it superseded two centuries worth of ground-based parallax and proper-motion data(!). The data also provided the foundations for a new reference frame—space astrometry has definitively replaced ground-based observations.

With the entire electromagnetic spectrum now opened for research, other information carriers are increasingly interesting. Apart from light of all wavelengths, there are also neutrinos and cosmic rays coming from space. Several large instruments have been developed over several decades to detect these elusive particles. In 2015, a fundamentally new kind of message was detected after forty years of exacting development work: a gravitational wave, probably produced by a pair of merging black holes. Just a few months earlier, the IAU had established a commission on gravitational wave physics—for once, the Union was ahead of the science! Several other detections have been made since then. In 2017, gravitational waves from a merging neutron-star binary was observed simultaneously by the LIGO-VIRGO gravitational-wave detectors, as well as by telescopes at every conceivable wavelength of the electromagnetic spectrum. This marked the addition of gravitational waves to the astronomical toolbox—another significant step in the direction of 'multi-messenger astronomy'.

The cosmic neutrino detector 'Ice Cube' on the Antarctic plateau has also added to the inventory of the astronomer's tool box—especially after it became possible to observe both up- and down-coming neutrinos. Since it is located in the southern hemisphere, the Earth acts to shield the background flux of up-coming cosmic rays coming from the North and separate the cosmic neutrinos, which are the key to Ice Cube point source searches.

But observations are no longer the only way to get new data. The rapid development of supercomputers have made large-scale modelling of astronomical and even cosmological processes possible, shedding light on phenomena ranging from supernovas to galaxy cluster formation and giving rise to a new discipline: computational astrophysics.

The Pluto 'Affair'

Astronomy has always been one of the most public of sciences, and popularisers from Camille Flammarion via Carl Sagan to Brian Cox have reached audiences of millions. The IAU had previously had no role in this, however, except for local outreach events during General Assemblies. Most of the educational activities were aimed at specific audiences: astronomy students or science teachers. Public outreach was an activity of individual astronomers, or at most national committees. The NEO and Pluto debates in the 1990s had demonstrated that the IAU had a public role to play, however. Public interest in astronomy called for a public authority to speak on behalf of the professional community. This happened again in 2006, albeit in a way that the IAU was not prepared for.

Definitions of scientific terms usually attract little attention. Specialists may care whether 'planetary nebulae' are related to planets or not, or whether the order of star classification is alphabetical or the sequence O, B, A, F, G, K, M; these are hardly a matter of public interest. This changed when an apparently technical discussion on the scientific definition of 'planet' suddenly became a major public affair.[4]

The planet Pluto had been discovered by Clyde Tombaugh in 1930 as part of a dedicated search, organised and financed by Percival Lowell at his private observatory in Flagstaff, Arizona. Pluto was accepted as the ninth planet, even though it was much smaller and its eccentricity and inclination far larger than for any of the classical planets; its orbit actually takes it inside Neptune's orbit for a time. Therefore, some people in the field regarded it as a special kind of large minor planet, captured into its present orbit by a close passage of a major planet. By the 1990s, several other objects had been discovered at roughly similar or larger distances, as part of the 'Kuiper Belt', raising further questions about how unique Pluto was. Was it a planet or part of a larger group of objects? Clyde Tombaugh, the discoverer, defended Pluto's status as a planet until a few years before his death in 1997.

Brian Marsden, conversely, had for long suggested that Pluto resembled minor planets more than the other planets. In 1999, he even proposed to give Pluto an 'honorary' minor planet number: 10,000. Unexpectedly for both NASA and the IAU, the two main sponsors of the Minor Planet Center, this

---

[4] The Pluto affair has been described many times in popular books and journals, including by some of the people involved, such as DeGrasse Tyson (2009) and Brown (2010). A good overview of the affair and its background is provided by Dick (2013). Related documents and correspondence are stored in the IAU Archives box 35.

caused some public uproar, as already alluded to in the previous chapter. Was Pluto's status as a planet under threat? Many voices were raised. The IAU office was flooded with questions and e-mails, and with letters in 'defence' of Pluto. It turned out that especially school teachers and children had strong feelings about this issue. Planets and stars are popular among primary school age children (among scientific topics, only dinosaurs come close). Changing such a fundamental fact as the number of planets that children had learnt for sixty years was surely unheard of. Or perhaps not?

Remaining silent was not an option, so then General Secretary Johannes Andersen issued a press release stating that the IAU had *not* decided to assign Pluto a minor planet number, and that no official proposal had been made, even less approved, to change its status as the ninth planet. He added:

> From time to time, the IAU takes decisions and makes recommendations on issues concerning astronomical matters affecting other sciences or the public. Such decisions and recommendations are not enforceable by national or international law, but are accepted because they are rational and effective when applied in practice. It is therefore the policy of the IAU that its recommendations should rest on well-established scientific facts and be backed by a broad consensus in the community concerned. A decision on the status of Pluto that did not conform to this policy would have been ineffective and therefore meaningless. Suggestions that this was about to happen are based on incomplete understanding of the above.[5]

The matter did not go away completely, however. For example, the new Rose Center for Earth and Space in New York drew protests for treating Pluto as part of the Kuiper belt, and for not including a scale model as it did for the other eight planets; public discussions attracted hundreds of people. The issue was also debated at the IAU General Assembly in Manchester in 2000. But the matter really came to a head when new, so-called Trans-Neptunian Objects (TNOs) were discovered, some of which were more massive and more distant than Pluto. One of them, discovered by Mike Brown, was even larger than Pluto, and had a moon. This object was initially informally known as Xena; later it was officially, and fittingly, named Eris and its moon Dysnomia, after the goddess of discord and her daughter. Was this also a planet?

Perhaps surprisingly, there existed no official criterion to distinguish planets from minor planets and their moons—there had previously been no need for one. Around the same time, the discovery of massive exoplanets caused

---

[5] IAU press release 01/99, 2 March 1999.

similar questions at the other end of the scale: where was the boundary between a large planet and a small brown dwarf? The latter issue, however, has attracted much less public attention.

As the 1999 press release explained, providing a definition was clearly a task for the IAU. Therefore, the IAU established a working group of 19 scientists to formulate a scientific definition of a planet for submission to the General Assembly in Prague in 2006. It came up with several options, but opinions were so divided that it could not choose between them. In the end, three definitions were submitted to the Executive Committee, which then appointed a new, smaller committee.

Because this clearly was an issue with cultural and historical implications, it was chaired by the well-known astronomer and historian of astronomy Owen Gingerich, and it also included popular science writer Dava Sobel as well as IAU President-Elect Catherine Cesarsky. The committee met just before General Assembly. It proposed to define a planet as an object in orbit around a star that is in 'hydrostatic equilibrium', meaning that its gravity is strong enough to give it a stable spherical shape, and which is not a satellite. Pluto, which met these criteria, was to be the prototype of a 'pluton', a new class of planets, while Ceres, the largest asteroid in the main belt, which also met the criteria for planethood, would be a 'dwarf planet'. All in all, the Solar System would then comprise 12 planets in total (the classical eight plus Ceres, Pluto, Charon and Eris), with the prospect of finding more in the future.

The proposed definition would have to be approved by the IAU General Assembly in a resolution. Since it was a scientific matter, all individual members present could vote. (In 2003 this had been changed to voting by national representatives, after the 2000 General Assembly had approved a resolution against the advice of the Executive Committee, but after protests, this was revoked at the beginning of the General Assembly in 2006, just in time for the Pluto vote.)

Since it clearly was an issue that sparked public interest, and the Executive Committee wanted to prevent a media storm ahead of the General Assembly, it was not announced that the definition was on the agenda until the meeting had started. The resolution containing the proposed new definition was announced to the press, however, which published it before astronomers had a chance to see it. This caused quite some irritation. What happened next took almost everybody by surprise, including the committee, the IAU leadership, and the newly appointed press officer Lars Lindberg Christensen.

Normally, proposed resolutions at the General Assembly of the IAU raise few questions. General Assemblies—not the conferences, but the so-called business sessions in which administrative matters are addressed and resolutions and budget matters are adopted—are usually attended only by a small subset of the conference participants, including the nationally designated delegates and others interested in the administration of the Union. Until then, the participants had rarely voted against the advice of the Executive Committee, but this time was different. Many scientists were critical about the proposed definition, which opened the door to having dozens of planets in the solar system in the future. Also, researchers from different specialisations favoured different criteria. Planetary scientists who worked on orbits, for example, favoured including a 'dynamical' criterion: planets should be the 'dominant' objects in their orbit, ruling out objects that were part of asteroid belts.

The result was many debates in the corridors and improvised extra meetings throughout Prague. But what was even more surprising is that these debates were followed by hundreds of journalists and bloggers who descended on the conference centre, providing blow-by-blow accounts in newspapers, journals and on the internet.

The crucial planet definition vote at the IAU GA in Prague, 2006. Photo: IAU Archives/L. Holm Nielsen

It is important in this context to realise that although the interest was international, resistance to a potential 'demotion' of Pluto was especially strong in America. Pluto was seen as an 'American' planet: the only one discovered in America, by an American, as the result of the dogged perseverance of the proverbial 'man with a dream'—Percival Lowell, whose initials are immortalised in the name of the planet. The connection to the famous Disney cartoon dog may also have contributed to its popularity (the dog was named after the planet rather than the other way around: the planet was named after the Roman god of the underworld). Another group that opposed the change consisted of teachers and popularisers, who had to change the well-established list of nine planets. Finally, people involved in NASA's New Horizons mission to Pluto, which had just been launched, feared that any change of Pluto's status would undermine its political support.

The actual General Assembly session in which the decision would be taken was scheduled at the very end of the conference. About 400 scientists attended. The discussion on the relevant resolutions was chaired by Jocelyn Bell Burnell, the eminent astronomer and co-discoverer of pulsars, who was member of the resolutions committee. She led the difficult discussion in an incomparable way. Some changes to the resolutions were included on the spot. Someone even proposed to simplify the issue by only keeping footnote 1 to Resolution 5A![6] Finally, the resolutions were put to the vote.

Resolution 5A read:
The IAU therefore resolves that planets and other bodies in our Solar System be defined into three distinct categories in the following way:

1. A planet[1] is a celestial body that (a) is in orbit around the Sun, (b) has sufficient mass for its self-gravity to overcome rigid body forces so that it assumes a hydrostatic equilibrium (nearly round) shape, and (c) has cleared the neighbourhood around its orbit.
2. A dwarf planet is a celestial body that (a) is in orbit around the Sun, (b) has sufficient mass for its self-gravity to overcome rigid body forces so that it assumes a hydrostatic equilibrium (nearly round) shape[2], (c) has not cleared the neighbourhood around its orbit, and (d) is not a satellite.
3. All other objects[3] orbiting the Sun shall be referred to collectively as 'Small Solar System Bodies'.

Footnotes:

1. The eight planets are: Mercury, Venus, Earth, Mars, Jupiter, Saturn, Uranus, and Neptune.

---

[6] The session was recorded and can be seen at: https://www.iau.org/public/videos/detail/iau2006session2/

2. An IAU process will be established to assign the borderline objects into either dwarf planet or other categories.
3. These currently include most of the Solar System asteroids, most Trans-Neptunian Objects (TNOs), comets, and other small bodies.

This was adopted by a large majority of votes. But the sting was in a second vote, on just one word: Resolution 5B proposed to change 'planets' as defined in point 1 to 'classical planets'. Bell Burnell explained the implications using props, including a stuffed Pluto dog representing Pluto, a balloon representing the other planets, and an umbrella to indicate which of them would be covered by which term. If Resolution 5B was adopted, there would be two equal classes of planets, classical planets and dwarfs (stuffed dog and balloon both under the umbrella); if rejected, dwarf planets were not planets (balloon under the umbrella, but not the dog; some objected on purely linguistic grounds). But it was rejected by an overwhelming majority.

Resolution 6A stated that 'Pluto is a dwarf planet by the above definition and is recognized as the prototype of a new category of trans-Neptunian objects.' This was adopted. Finally, Resolution 6B which proposed the name 'plutonian objects' for this class was rejected however, leaving the naming for later. The term 'plutons' which had been proposed, was problematic because, in some languages, this was already the name of Pluto, and moreover it already was the geological name for a certain type of rock. The Executive Committee later decided on 'plutoids'. (The procedure by which this was done raised some eyebrows, but a renewed PR-disaster was narrowly avoided.)[7]

---

[7] IAU Archives box 35.

Not least the Pluto controversy at the IAU General Assembly in Prague in 2006 kept the press room busy. Seen to left and facing the camera is Brian Marsden of the Minor Planet Center, a frequently sought source for the journalists at this moment. Photo: IAU Archives/L. Holm Nielsen

According to Gingerich, Brian Marsden, who stepped down as the director of the Minor Planet Center during the same meeting, was 'quite entertained by the thought that both he and Pluto had been retired on the same day'.[8] Many other people were not so amused, however: the decision released an unprecedented barrage of protests. The IAU archive contains hundreds of letters, including many from children and even entire (mostly American) primary school classes, who regretted the 'demotion' of Pluto.[9] Bumper stickers and internet memes abounded, the American Dialect Society chose 'Plutoed' (meaning demoted, devalued) as 'word of the year 2006', and the Congress of the US state of New Mexico voted that Pluto was still a planet as it passed overhead.

Most of the reactions were either disappointed or comical (or both), but some were sharper, including accusations of 'Anti-Americanism'. For the first time, the IAU's authority in astronomical issues was questioned. Even though, as the 1999 press release stated, this authority was not written in any law, it had never really been challenged before. Now it had been by some, notably

---

[8] Quoted in Gingerich (2010).
[9] IAU Archives box 35.

including the principal investigator of the New Horizons mission, who founded a company that sells names of exoplanets and Mars surface features to the public.[10]

Within the scientific community, the IAU's authority was not questioned, however, although some questions were raised about the fact that such a small number of scientists—the 400 people who were present in the room—could decide on such a wide-ranging issue. This boosted a discussion on electronic voting, which was finally introduced in 2012. From then on, all IAU members could participate in voting, whether they were at the General Assembly or not.

Many astronomers were baffled by the whole affair. Some astronomers still do not understand the fuss about names—not scientific facts—concerning what has been called the 'least important part of the Universe'. Over time, however, insight changes, and the equivalent had even happened before: Ceres, Juno, Pallas and Vesta been generally counted as planets in the early nineteenth century; only when more, similar objects were discovered after 1845 were they reclassified as asteroids or 'minor planets'—although as we saw, a precise definition of the difference was not formulated until 2006. Despite the new definition, Pluto still got its minor planet number, since a new classification system for dwarf planets has not been developed. It is now minor planet no. 134340.

In the end, the 'Pluto affair' was a major catalyst for the IAU's public role, despite the fact that the amount of public interest took everybody by surprise. The result was that many more people heard about the IAU—albeit mostly negatively—than ever before. The authority of the IAU in astronomical matters, which had never really been challenged before, was sorely tested. This was difficult for an organisation that had always stayed out of public disputes, but any damage was not permanent.

The amount of public interest also proved the huge popular appeal of astronomy. As many astronomers and educators already knew, stars and planets are an excellent way to start talking about science, and clearly people care about names. After 2006, the IAU has organised several carefully crafted public naming campaigns for exoplanets and for surface features on Pluto—if only to demonstrate that astronomers still care about Pluto, whether it is a dwarf planet or not.

---

[10] See www.uwingu.com.

Ron Ekers, IAU President and Internationalist

Chair of the Working Group on Future Large Scale Facilities (1998–2000), President Elect (2000–2003), President (2003–2006)

Interview conducted in Vienna on 27 August 2018

For a boy born on a farm in rural South Australia, even with an early interest in science, it wasn't obvious that he would end up as a director of the VLA, an important—in many ways transformational—research facility in the USA, then as director of the Australia Telescope National Facility and as President of the union of the world's astronomers. But as many astronomers will attest: Anything can happen in astronomy.

Ron Ekers at the IAU General Assembly in Vienna. Photo: C. Madsen

Ron Ekers graduated from the University of Adelaide in 1963 and obtained his PhD in astronomy at the Australian National University (ANU) in Canberra in 1967.

'I joined the IAU in 1971 and simply participated as an ordinary member to various meetings. Yet, I had been more deeply involved in URSI and I had been elected President of the URSI Commission J on Radio Astronomy in 1987, so I had some experience in the ways and work of international scientific unions.

In 1998, when I had just finished being the director of the VLA observatory in New Mexico and on my way to taking up the position as director of the Australia Telescope, I received a letter from the IAU president, Bob Kraft, asking me, if I might be willing to serve the IAU as President, first, of course, becoming President-Elect. He noted in his letter that I would be only the second radio astronomer to assume this task, the first one being Robert Hanbury Brown, also from Australia. Indeed Radio Astronomy had been a bit of an orphan with respect to the IAU, and it was Joe Pawsey, leader of the first radio astronomy group in Australia, who worked to incorporate it into the IAU and became president of Commission 40 in 1952. I don't think that the IAU was hugely enthusiastic about this strange new field of astronomy at the time but the links to traditional astronomy were very important for the radio engineers. In retrospect, though, you could say that

(Continued)

it was the first step towards multi-messenger astronomy. Today, however, we have a good coordination of roles between URSI and the IAU.

So I became President-Elect at Manchester in 2000. Manchester appeared to me to be a difficult General Assembly. Only five symposia had been proposed, participation was modest and the lack of formal agreement between the Union and the host country created difficulties. With Sydney having already been selected as the venue for the next GA, I was concerned. The issue of participation numbers with the ensuing financial consequences worried me.

In any event, I went to see Lodewijk Woltjer, who had been IAU President and had initiated quite a few changes including the introduction of the Division structure, to understand the reasons for those changes. As Woltjer told me: "Any changes will take six years to implement, so if you wish to make changes, you'd better start immediately". But I was "only" the President-Elect …

One important point made by Woltjer was that the new Division structure was primarily to improve the upwards communication from the members of the Commissions to the Executive Committee, which had been sorely missing. It seemed to me that this was a democratic problem—the members wanted to be more involved, but the established structures did not facilitate this.'

The disconnect came to a head at the Manchester GA when the resolution addressing the need to preserve the scientific data from more than a hundred years of photographic plates was approved by a majority of the rank and file members—but against the advice of the EC. Members were revolting against being told how to vote.

'This was unusual and it did not go down well with the Executive Committee. However, since at the same time the statutes and by-laws were undergoing a revision, a change was included to rescind the right of individual members to vote on scientific matters. I must add that most of the revision of the rules were justified and very sensible. This change was not noticed by the IAU members. The draft rules including this change had been sent to the National Committees ahead of the GA without any objections being raised, but after the Sydney GA, there was a strong reaction from the membership objecting to the lack of any democratic process. This incident also showed that the National Committees were not consulting with their local astronomy communities before they voted at a GA so this aspect of representation was dysfunctional. It was clear to me that we needed to reinstate the voting right for individual members, but it could at the earliest happen at the following GA—the Prague 2006 meeting.'

A related issue in terms of improving the internal functioning of the IAU was to breathe life into the new Division structure by devolving scientific responsibilities, e.g. in setting up the scientific programmes to the Division Presidents. Up until now this had been the domain of the Executive Committee and the Vice-Presidents.

(Continued)

The IAU Executive Meeting 76 in St. Petersburg in May 2002. This was the first time the EC met with the Division Presidents. Photo: Collection Ron Ekers

'I put quite a bit of effort into making that change. For the EC meeting in St. Petersburg in 2002, we for the first time invited the Division Presidents to attend. We had a lively and very constructive discussion and it gave us in the EC real feedback. At the following EC meeting, EC80 in Rome, we took the next step in that the Division Presidents, who were, of course, the real experts within their specific fields, were tasked with selecting the scientific symposia instead of the EC. While all of this was positive, it had an impact on the budget, mainly in terms of reallocating resources. That meant that some inevitably would lose funding and influence—the Commissions. Understandably, many Commission Presidents were unhappy about this and in the end, despite my own conviction that reforms were needed within the IAU, I found it wise to slow down the speed of change to give people time to adjust. This was also dictated by the need to be listening to the members and not simply dictating changes.'

With these initiatives and a proactive involvement of the host country, Ekers' initial worries about the Sydney GA were put to bed. The number of symposia proposals increased five-fold, thus allowing for a critical, quality-based selection. Also, the number of attendees increased again to more than 2000, which made the meeting financially viable. Sponsor funding came in handily, not the least because the opening ceremony took place in the famous Sydney Opera House providing a memorable event.

(Continued)

An IAU General Assembly comes with a cost. Next to the conference fees, sponsors are needed, including companies and organisations involved in astronomy. Many of them put up information stands in the conference exhibition area, like here at the 2003 Sydney GA. Photo: ESO Archive/E. Janssen

'We had noticed that there were people directly involved in supporting astronomy who would not necessarily meet the strict criteria for membership, but who could still play an important role in the Union—educators, librarians, historians being typical examples, and with the changes of the statutes and by-laws at Sydney—strongly influenced by Bob Williams and Johannes Andersen—this now became possible. However, it still left out another group: The young astronomers. The notion was that they would be "looked after" by the relevant national committees, but in this day and age, early-stage researchers move around, typically from country to country, and therefore do not have strong links to any national committee. We did not have a good way to identify them, but we began to invite young astronomers and have gatherings for them at the GA. In any case, we started the process of engaging young astronomers which has now—15 years later—come to full fruition with the new membership category of junior astronomers.'

Given the trend in the second half of the twentieth century astronomy to converge around large and expensive observational facilities funded by international consortia, it was natural to ask the question of the possible role of the IAU in these developments.

'At the 1994 GA in The Hague, a session on future large-scale infrastructures was organised. After that Francoise Praderie, Harvey Butcher and I suggested to establish a working group on this topic under the IAU EC and it remained active for the next nine years. The working group was then moved into Division B under the latest reorganisation, and quite a lot of corporate memory disappeared.

(Continued)

At the 2018 GA in Vienna, the working group under the EC was relaunched by the incoming President. This is good, but it points to a problem of how to retain corporate memory in an organisation with a rotating executive and a very small permanent secretariat in Paris. I accept the value of a "lean administration", but I believe the IAU administration is too lean for an organisation of this size. Just compare it with the other Scientific Unions or the office of, say, the American Astronomical Society.'

Ekers' term as President ended at the IAU GA in Prague. Key topics at Prague included the voting issue and the planet definition proposal and, of course, this GA would also see the IAU beginning intense preparations for the planned International Year of Astronomy. As regards the latter, unbeknownst to the members, dark political clouds were gathering at the level of the United Nations.

'During a meeting with the Australian and Italian ambassadors at the UN in New York, it became clear to me that the road ahead would be bumpy and that the potential role of UNESCO had been overestimated. This would have to be addressed by my successor as President, Catherine Cesarsky.'

'As regards the voting question, the EC79 meeting in Mexico City, we had an impromptu discussion about the functioning of National Committees. We realised that no operating guidelines for the National Committees existed, which meant huge differences in the way the NCs worked in each country. It had therefore also become clear that expecting the National Committees to collate the views of the membership and that the NC votes at the GA would thus reflect these views was too idealistic. The need for reinstating the voting right for individual members was therefore obvious and it had to done already during the first business session at Prague so that members could vote in the second business session.

The discussion about voting continued at the subsequent EC meeting in Rome in 2005. The issue of electronic voting was broached, but I was not in favour at that time. I thought that with electronic voting the very opinionated people would make a point of voting, while others would not. Also members not present at a GA would not have been exposed to an open discussion. By contrast, keeping the vote to those attending the GA meant presenting the proposal to a presumably unbiased audience who as knowledgeable astronomers would be exposed to the arguments and could thus make an informed choice, much as it's done in a parliament. To facilitate this, at the Sydney GA, the people proposing a resolution would be given the opportunity to explain the rationale, rather than simply reading the resolution text, which for example in the case of some of the astrometry resolutions are quite complex. Even if the vote would only comprise a subset of the entire membership, it would ensure that the issue in question had been thoroughly considered as befits a scientific matter.'

Be that as it may. The test would come in Prague.

'In 2005 the Working Group on Nomenclature faced a question of naming relating to the naming of 2003 UB313 which was a Trans Neptunian object discovered by Mike Brown and estimated to be larger than Pluto. The question arose because the naming conventions differ between planets and minor bodies in the Solar System. Following normal IAU procedures we asked the Planetary Division to make a recommendation to the EC. In November 2005, after two attempts by Divisional working groups, no decisive recommendation had been made. The working group had agreed on the science and the evidence for more

(Continued)

than one class of solar system object but they could not agree on a definition of a planet and presented the EC with three options, either (a) set Pluto as the minimum size, (b) include the dynamics of the system, or (c) use round shape, i.e. hydrostatic equilibrium as the decisive criterion.

In March 2006, the Executive Committee set up an interdisciplinary working group to deal with the matter. The group was to include history, science outreach, writing and education as well as planetary science. A condition for some of the working group members was that they could work "in peace", i.e. without the risk of being approached by outsiders including the media trying to influence or publicise their activities. The working group came up with a recommendation, and the EC had to decide on how to present it to the IAU members.

It should be noted that naming is not a science in itself, but it's part of the intellectual—linguistic, in fact—infrastructure that we need to do science. Also, the IAU neither has the authority to make laws nor to enforce their implementation. Naming is simply a consensus exercise. Given our awareness of the sensitivity and the risk of derailing the process, we made the decision to not announce anything about it before the first business meeting of the GA in Prague. To be honest, to this day it bothers me because even if it probably was the right thing to do in this exceptional case, it ran counter to my own ideas about strengthening the involvement of the membership and conducting all IAU business in an open and transparent manner.

We knew that once we made it public, there would be strong media interest and so we appointed a press officer, Lars Lindberg Christensen from ESO. I believe this was the first press officer in the history of the IAU. Lars had been briefed in advance of the Prague GA. As expected, when the first press release went out, indeed the press started pouring in and were invited to attend all discussions. I have to say that whilst the preparations had happened behind the scenes, from this stage on, everything was played out transparently and in full public view.'

Ron Ekers addressing the participants at the IAU GA in Prague. Photo: Collection R. Ekers

(Continued)

The first step, however, was the change of the rules to reinstate the voting rights of the individual members, which happened during the first business session. Now the stage was set for an intense discussion and final decision by the IAU members in attendance.

'At a special business session, we took a straw vote on the initial resolution and it became clear that the pure version of the proposed text would not pass. Aside from the realisation that the planet definition would not simply impact on the naming of "new" objects, but also potentially affect the status of Pluto—of which different opinions were held—the question basically pitched the "dynamicists" against the planetary geologists. And since the GA also contained a symposium on Near-Earth Objects, it meant that there was a strong representation of dynamicists at this GA.

We obviously had to have a finished resolution text ready for the final business session of the GA. That put a lot of strain on the resolutions committee but in the end, a new text was put together. The resolution committee was chaired by Chris Corbally and the resolution was presented and explained by Jocelyn Bell Burnell. They did an outstanding job under very difficult circumstances. To make matters simple, we split the resolution into parts before the final vote. With the first vote which included a planet both being round and clearing it's orbit passing comfortably, the second vote on whether dwarf planets were also planets was now a critical step if Pluto was to retain its status. It failed so Pluto had now de facto been demoted. The third vote that was about putting Pluto into a special class, initially to be called Plutons, created new problems due to protests from the international geology community for whom "Pluton" has a different meeting. Because of the secrecy before Prague, it had not been possible to discuss this with other relevant scientific unions. The naming suggestion was changed on the fly to "Plutonians", but this failed and the decision on the new name for this class of object was postponed.

The planet definition vote again raised the question of electronic voting since only the small fraction of the IAU members present in Prague could vote. In retrospect, I don't think that an electronic vote would have helped at all, because the outcome was the result of a serious open discussion, occasionally fierce but by and large based on scientific arguments, and that would have been very difficult to do this with a purely electronic discussion and voting process.'

The Pluto controversy made waves in the international media, especially in the USA, though much less among the scientist themselves.

'There was a group of colleagues who thought we had been very short-sighted by not addressing the issue of extra-solar planets. But the fact is that we did this on purpose. It was already a very difficult to obtain a clear decision and so we wanted to keep the process as simple as possible. But it is also true that many professional astronomers, perhaps even the majority, didn't really care.'

Indeed, its vivid discussions and public resonance notwithstanding, many IAU members neither considered the Pluto case interesting nor important. For some, it was more like a storm in teacup.

'It generated enormous visibility for the IAU, as it set out to do what it's supposed to do and there are still repercussions. It had perhaps had more influence than anything else the IAU has ever done. So, not exactly a storm in a teacup!'

The Prague GA stands out as one of the most memorable GAs in recent times. To the President who oversaw the proceedings at this meeting but also in the

(Continued)

100-year long perspective of the IAU, how would he summarise the importance of the IAU?

'I see the IAU as an opportunity to promote international science and to promote the use of science as a mechanism for international collaboration. That goes from simply meeting people to the realisation that organisations like the IAU are often better at international interactions than governments. Official governments are actually often surprised by the ease with which scientists can set up international collaborative activities. In this sense, the international community of astronomers, the body membership of the IAU make an important contribution to the world that goes beyond the mere development of our own science.'

Jocelyn Bell Burnell: The Right Person in the Right Place at the Right Time

Chair of Resolutions Committee (2006–2009)

Interview conducted in Vienna on 26 August 2018

It would appear to be a lucky coincidence, but sometimes someone can be the right person in the right place and at the right time. In the case of Jocelyn Bell Burnell, this happened more than once. In 2006, this was a gift to the IAU.

Jocelyn Bell-Burnell at the IAU General Assembly in Vienna. Photo: C. Madsen

Jocelyn Bell Burnell was born in Belfast, Northern Ireland, in 1943.

'I grew up in a small country town. I started doing science in secondary school at the age of 12, but as was common then—it was in 1955—the boys went to science classes while the girls were sent to "domestic science"—cookery and housekeeping. I wasn't happy and neither were my parents. They spoke to the head teacher and he gave in. Next time the science class met, there were three

(Continued)

girls and all the boys. We did physics in the first term and I came out the top of the class. The only thing I got wrong was the speed of light. I wrote down the correct number, but when I looked at it again, I thought this was a very big number, so I changed seconds into hours! Through my teenage years, I was wondering what kind of physicist I might become. My father brought me two books: Fred Hoyle's "Frontiers of the Universe", which I understood, and "The Unity of the Universe" by Dennis Sciama, which I did not fully understand. So I did a physics degree at the University of Glasgow—as the only female in the honour's class.

Meanwhile, I'd decided that wanted to become a radio astronomer since they don't have to work at night! I was prepared to go to Australia, but I had also put in an application to Cambridge, and to my surprise they accepted me. As I was interested also in geophysics, I was allowed to choose to work on an interplanetary scintillation project, the purpose of which was to build a radio telescope to find compact radio sources that were suspected to be quasars—of course, a hot topic at the time.

I spent two years with a small team on building the radio telescope. When it was finished, the rest of the team moved on to other projects but I remained, becoming responsible for operating the telescope, while now being in the last months of my time as a postgraduate student.'

Bell Burnell achieved world-fame when in 1967, observing with this telescope, she made the discovery of PSR B1919+21, a previously unknown type of object that subsequently became known as pulsars. The story of the discovery has been told elsewhere, but in essence it is a fascinating tale and fine example of two of the most remarkable elements of scientific work—on the one hand the dark horse of scientific discoveries: serendipity and, on the other hand, meticulous work and incredible stamina once a hint, however feeble, of 'something interesting' has been registered in a curious mind. In 1974, the discovery of pulsars led to the awarding of a Nobel prize, albeit not to Bell Burnell. Once again, much has been said and written about this, though notably, she herself has consistently downplayed it.

'At the time, it was Fred Hoyle who stirred that pot. For me, the issue was as follows: I was then a post-doc, married to somebody who kept moving around. I also had a small child, and given this personal situation for me staying in astronomy was a major challenge. So I needed references and stirring the pot is not helpful in this situation. Furthermore, this was the first time that the Nobel Physics Committee had recognised anything in astronomy. We had now finally convinced the physicists that there was good physics in astrophysics and I recognised that it had created a precedence and that other astronomers would likely go through that door. Also, I probably wasn't prepared for the fuss about "Jocelyn not getting the Nobel Prize". To my colleagues in the UK, it became known as the "No-Bell prize".'

Might gender issues have been at play here?

'Yes. It was difficult to be a female scientist. I made the mistake of wearing my engagement ring into the department not realising that it sent out the signal that I was leaving because, in Britain, married women did not work. On top of that, my husband did move job to different parts of the country and I found myself, again and again, writing a begging letter to the nearest astronomy institute director to get even a part-time job. The advantage, however, was that I became familiar with gamma-, X-ray and infrared astronomy, even mm-wave

(Continued)

astronomy. Later, with a divorce behind me, I was now free to go where I wanted and I was appointed as a physics professor at the Open University in Milton Keynes, studying energetic binaries—at whatever wavelength we needed.'

'My first encounter with the IAU was at the General Assembly in Brighton. At the time, I was working in gamma-ray astronomy at Southampton University and the gamma-ray people were not really "switched on to the IAU". I was probably the only person from Southampton at the meeting and it felt big—and with a lot of big names. And clearly, I got the bug! I have been to quite a few of the GAs since. I wasn't aware of how the IAU conducted its business—I was there for the science. I was already well used to being the only female—in fact since my under-graduate studies. So going to a meeting so overly dominated by men as the IAU GA at the time was nothing unusual for me. I didn't particularly like it, but that was simply the way it was … I guess it took some courage, but a lot of my life, as a woman I had to be courageous and go into an event or a big lecture. Still, I was not really involved in the early struggle within the IAU to improve the lot of female scientists. Instead, during my later career, I have focused on the situation in the UK academe with the Athena Swan initiative. But at the IAU, I wish to acknowledge the efforts by Johannes Andersen when he was General Secretary. The IAU had data on membership and he collected the data disaggregated by sex, which provided us with very useful information in that respect—it was pio-neering work.'

Then came Prague.

'In 2006, I was on the resolutions committee, though not yet in the chair. When I arrived in Prague, I had somehow picked up that there was something going on regarding Pluto, but the Executive Committee had clearly run a very tight ship with very few people in the know. It happened, however, that an announcement was sent out to the media before the astronomers received the information. One of the first indications to us was when Birgitta Nordström received a telephone call from a Danish newspaper asking what was going on. Then the media bar-rage started. This led to a strong reaction on the part of the IAU members, who felt that they should have been first to receive the information.'

When the resolution was presented, it happened in the usual way with the key people in an elevated position on a stage and behind a desk. Presenting the proposal and the rationale took a while, so in the end, there was little time for a real discussion. If the issue was contentious, even if unintended, the choreogra-phy did not help. That set the scene for a clash that had never been seen before in the IAU.

'People were standing in line to speak and several aired their frustration about not having been informed in advance, including Chairpersons who felt they should have been privy to advance notification. That, of course, included the planetary scientists from Division III. In the end, a straw vote was taken and the proposal by the Executive was rejected.

After the stormy session, the EC held an emergency session and they called in the chair of the resolutions committee and myself. I made various suggestions, but coming back into the room after a short break, I was told: "Jocelyn, you're fronting a meeting this evening!"

The organisers got a room for 150 people, but about 250 turned up, so many had to stand up. All I had to do, I thought, was to conduct the meeting in a dif-ferent way than the lunchtime meeting. I took care, not to have any physical

(Continued)

barriers and to let people speak. I did limit interventions to two-three minutes, but basically let people express themselves, partly how they felt, but of course also what they thought should now happen. In particular, the dynamicists were unhappy with the proposed resolution. But it became clearer at that meeting what might work and what not. The members of Executive were not happy, either. Undoubtedly wary of the societal aspects, they were keen to retain the thrust of the original resolution text, which would have allowed to keep Pluto as a planet. So I ended up doing shuttle diplomacy between the various antagonists.'

But it would take more than one meeting to sort out the issue and so, the discussions continued during the conference days. The Chair of the Resolutions Committee and I worked together in formulating a new proposal.

'In the end, we succeeded! A couple of days before the second business session, we had a new text that was then published in The Nuncius, the conference newspaper. But it wasn't over yet. I went to very few of the science meetings. Instead, I sat in a spare booth in the exhibition area where people kept seeing me with additional suggestions for the wording.'

It was decided that Bell Burnell would also front that part of the second, and final, business meeting, conducting the discussion on the Planet Definition Resolution. It was to become perhaps the most memorable event in the history of the Union.

'I have given many talks and presentations in my life, but this time I was really scared whether I would do it well, whether it would be clear enough for everybody to follow.

Perhaps the most spectacular Pluto show ever—Jocelyn Bell Burnell introducing the vote about the definition of a planet and the status of Pluto. Photo: IAU Archives

(Continued)

Before the final meeting, i.e. the second business session, I went shopping in Prague. I managed to find a child's balloon, a package of cereals (to illustrate the asteroids) and a lemon to illustrate the non-spherical bodies in the Solar System. I also got a toy dog from the press office and an umbrella. It was partly to ensure that the reporters would understand, but what I hadn't realised was that after voting on the first resolution—i.e. before I came to use these visual aides—a lot of people left the room. It was the press people, who rushed out. We had "killed" Pluto and they were out to report it. I'm glad I didn't initially realise how many reporters there were and that the CNN and BBC were live-streaming it. The interest was simply huge.

Jocelyn Bell Burnell introducing the GA proceedings at the GA in Beijing. Photo: C. Madsen

Six years later, at the Beijing GA, I was invited to become part of the small group of IAU people to meet with the then Vice-President Xi Jing-Ping and his entourage. After the formal opening, I was on stage to give an introductory talk. I tried to frame it in a historical context and incorporate as many international aspects as possible. Somewhat later I had the pleasure along with George Ellis of being on the review panel for the Office of Astronomy for Development.'

By 2006, Bell Burnell had concluded a stint as Dean of Science at the University of Bath (2001–2004) but remained Visiting Professor of Astrophysics at Oxford.

'There are advantages of not receiving the Nobel Prize. If you get a Nobel Prize, there's a fantastic week in Stockholm and then nobody gives you anything else because they do not feel they can match the Nobel Prize. If you do not get

(Continued)

the Nobel Prize you get every other prize you can think of. And a good party now and then! Much better!'

Among the many honours and distinctions accorded to her are the Albert A Michelson medal (1973), the J Robert Oppenheimer Memorial Prize (1978), the Herschel Medal of the Royal Astronomical Society (1989) and the Special Breakthrough Prize in Fundamental Physics (2018). In 2007 she received the title of Dame Commander of the Order of the British Empire.

Catherine Cesarsky: No Quiet Days

President of Commission 48 High-Energy Astrophysics (1985–1988), Vice-President (1997–2003), President Elect (2003–2006), President (2006–2009)

Interview conducted in Paris on 30 June 2017

There was no shortage of issues or events during Catherine Cesarsky's time at the helm of the IAU, but it marked the start of a transition of the Union from the ways of its long past to a Union with a strong public profile.

Catherine Cesarsky at the prestigious Institut de France, home of the Académie des sciences. Photo: C. Madsen

Catherine Cesarsky was born in France. Her family moved from France to Argentina in 1945, after the war, and after her schooling mostly in French, in 1960, she began her studies in Physics at the University of Buenos Aires. Cesarsky obtained a PhD in astronomy at Harvard University in 1971 and subsequently worked at CALTECH. In 1974, she returned to France to work at Service d'Astrophysique of the French Atomic Energy Commission.

'The IAU did not play an important role in my life as a young researcher. In my field—I worked on the propagation and origin of galactic cosmic rays—COSPAR was seen as the primary place to go to. "My first" IAU General Assembly was the one in Grenoble in 1976. It was about two years after my return to France. I

(Continued)

remember this meeting as a nice opportunity to meet many interesting people, including Vitaly Ginzburg, whom I had not met before.

Franco Pacini, who of course had long-standing relations with the IAU, has always been an enormous help for me. I had met him briefly in America and as soon as I arrived in Europe, he involved me in as many things as possible. In 1982, I was asked to play a role in the IAU Commission 48 for High Energy, which was separate from the Commission for Space. At the time, the Commission was chaired by Riccardo Giacconi and I became its Vice-President. High Energy in the IAU "was very small", but we managed to get a nice topical meeting in connection with the General Assembly in New Delhi, and again in Baltimore in 1988 where I was Commission President.

Later, my involvement with the IAU deepened. Around the time when Jorge Sahade was IAU President, he asked me if I would consider becoming IAU General Secretary, but since I was already director of a laboratory at CEA and Principal Investigator of ISOCAM, I declined.'

But in 1997 Cesarsky became IAU Vice-President for two periods. This was the beginning of an uninterrupted 15 years of tenure at the IAU Executive Committee.

'In 2003, I became President-Elect and thus set to become President three years later. As IAU President at the time, the special nominating committee was headed by Franco Pacini. So it was Franco, once again, who supported me. He called me about becoming IAU President. I was Director General of ESO at the time, but I agreed with the knowledge that the initial period of three years would entail a workload that would be manageable vis-à-vis my "day-job". After all, serving the IAU is really a service to the entire community.'

The tenure of Cesarsky was a period of great changes at the IAU, with conspicuous way-points: The Great Pluto Debate marked a desire to finally solve not only the long-standing issue of the status of Pluto itself but also the underlying nomenclature and principles behind the naming of celestial objects; The International Year of Astronomy, pronounced by the United Nations, was a unique chance for world astronomy to enhance its visibility and for the IAU, leading the effort, to develop projects with lasting impact; and, internally, once again tackling programmatic and house-keeping problems, from the division structure to fundamental administrative matters.

'In 2003, Ron Ekers, someone I'd known for almost all my professional life, became President of the Union, leading it for the following three years. Two topics during that period stand out: The issue of planet definition and the early preparations for the IYA.

(Continued)

The 2006 planet definition working group: André Brahic, Iwan Williams, Junichi Watanabe, Richard Binzel, Catherine Cesarsky, Dava Sobel and Owen Gingerich. Photo: IAU Archives

As regards the question about Pluto, following the discovery of transneptunian objects with properties akin to Pluto, a working group under the responsible Commission had struggled with the issue for some time. But Ron rightly saw that it had repercussions beyond the world of astronomy, so he appointed a more diverse committee to tackle the matter, including even a well-known science writer, Dava Sobel. The committee, named the Planet Definition Committee, was chaired by Owen Gingerich and it met at the Observatoire de Paris on 30 June and 1 July, 2006. I came to the meeting as an observer on behalf of the Executive Committee. Although I did not feel too strongly about it, my own view was that Pluto should not retain its status as a planet. But on the whole, the committee wished otherwise, meaning that not only Pluto but also the "new" objects would become planets. This meant that the Solar System would have "grown" to have 12 planets, with" Xena" (2003 UB$_{313}$) as its first new member. I admit that I left the meeting with an uneasy feeling, but now there was a proposal to present as a draft resolution to the IAU membership at the General Assembly in Prague.

Alas, the preparatory work had been done under time pressure, and also at the General Assembly, timing turned out to be a problem. On the first day of the General Assembly, the draft resolution based on the recommendations of the Gingerich Committee was sent to the media, despite my misgivings. This was in recognition of the strong public echo to be expected, but it did not go down well among some astronomers, including the planetary scientists of Division III present at the meeting, who only learnt about it on the following day. What ensued were numerous discussions during the 12 days of the Prague meeting. Given the commotion, I encouraged those who were unhappy with the original

(Continued)

resolution to formulate alternative proposals. In the end, it resulted in a revised draft resolution that was put before the members during the second session of the General Assembly on the last day. Now, according to the new proposal, Pluto would be accorded the status of a dwarf planet—and the Solar System would thus contain eight planets, and a number of dwarf planets, three at the time, which was expected to grow steadily as observations of the Kuiper belt objects intensified. I voted in favour of the final resolution, demoting Pluto.

The following morning we had a press conference, and now it was me, as the new IAU President, who had to defend the decision in front of the media, including a very critical US press.

For quite a while I received enormous amounts of messages and phone calls from journalists from all over the world. Resuming my "normal" life as Director General of ESO, which implied almost constant travelling, I found myself speaking to many a journalist in the corridors of quite a few airports in those days. For the French media, I was helped by André Brahic, whereas I looked after the English and Spanish-language media.

I realised soon that the Pluto debate, which had not, I admit, been very well managed at IAU, had triggered a huge amount of enlightening discussions in schools all over the world, as children could see a patent example of the way new discoveries lead to new concepts, and scientific knowledge evolves.

From an astrophysical point of view the decision to "demote" Pluto did not reduce the importance of this object. In 2008, the Executive Committee, after consultation with the two nomenclature committees and with Division III, decided to use the term plutoid to designate transneptunian dwarf planets similar to Pluto. The message we wished to convey was that Pluto should be seen as the prototype of a very interesting and important "new" type of object. (At the time, Eris was the only certain plutoid other than Pluto itself, but now there are at least two more, and a long list of candidates). I was keen to do that, just as I later, from the HST STAC, also supported the search with the HST for an object which could be approached by the New Horizons Mission. Still, many people in the USA, especially the group around the New Horizons Mission, and Owen Gingerich himself have kept vocal about this issue, insisting on "reinstating" Pluto as the ninth planet of the Solar System.'

In parallel to the lively, occasionally bitter, discussions about Pluto, Cesarsky's term as IAU President was, of course, clearly associated with the preparations and implementation of the IYA2009.

(Continued)

'The IYA was really Franco Pacini's brainchild, and I found it a fantastic idea. In 2003, this resulted in the formal resolution proposal presented to the IAU General Assembly, which was, of course, enthusiastically received. And so, we started serious discussions about 'The Year' already when I was Vice-President.'

Pacini obviously took the lead, keen to make a success of it, but partly because of his beginning illness, partly because of the sheer magnitude of the task at hand, increasingly this became a heavy-lifting exercise needing many strong shoulders.

Catherine Cesarsky and Franco Pacini during the EC meeting in Rome in April 2005. Photo: Collection Hans Rickman

'After Sydney, we agreed to put in place a dedicated working group to deal with the IYA, but for a while, little happened. And as Franco became increasingly frail, his involvement became less. This was undoubtedly difficult for him and—not the least given my own personal relations with him—I was sad about it, too.

(Continued)

In any event, before the Prague General Assembly, it was decided that both Ron Ekers and I should join the committee. After "Prague", I took the chair of the committee and put in as much time as I could into it, working with Lars Christensen, Pedro Russo and their team, for the contents, and with Claus Madsen and Enikö Patkós for the political lobbying. During the Year itself, a large number of events and activities took place. The main official event was the grand opening, held at UNESCO in Paris in January 2009. Of course, the relations between the IAU and UNESCO go back a long time and I especially appreciated working with Anna Sidorenko, the UNESCO programme specialist for World Heritage sites. To mention other events: A European opening ceremony in Prague in front of the astronomical clock in the presence of the European Commissioner for Science, the IAU General Assembly itself, taking place in Rio de Janeiro in August, and finally, the closing ceremony, fittingly held in the historical Aula Magna of the University of Padova. At this ceremony, it was gratifying to see Franco again and to listen to his speech. It gave us an excellent opportunity to show our great appreciation of him in his home country.'

UNESCO Director General, Mr. Koichi Matsuura, addressing the invited audience at the official IYA Opening Ceremony. Photo: IAU Archives/J.F. Delgado

Whilst the IYA came to an end in early 2010, another important IAU initiative was started, embodied in the Strategic Plan, mainly elaborated by George Miley. This entailed the allotment of funds on a scale not previously done by the IAU in favour of astronomy development in the developing world, and that at a time when the IAU finances appeared to be under considerable strain.

'Shortly after "Prague", we became aware of the fraud issue at the Paris Office. I was obviously extremely upset, as was Karel van der Hucht, the General Secretary at the time. The problem had been uncovered by Vivien Reuter, the new Executive Assistant, who did a fantastic job to document it thoroughly.

(Continued)

Vivien was really a literary person and yet she dug into the nitty-gritty and technical details about the accounting. She was amazing.'

In these dark days, however, the IAU was blessed with having a strong leadership team in place with highly experienced people—Cesarsky as President, van der Hucht as a meticulous General Secretary, Ian Corbett—himself with strong managerial experience—as Assistant General Secretary and Bob Williams as President-Elect.

'Karel had done a great job, country by country, in convincing them to raise their contributions, and the flip side of the Orine affair was that we realised that our basic finances were in a better shape than had been thought. We also recovered some of the money and that made it easier to allocate the funds in support of the Strategic Plan. In any case, openness about the situation was the only way forward, and so it was addressed at the General Assembly in Rio.

This meeting was supposed to be the highlight of the IYA2009. There had been the expectation that the Opening Ceremony would be attended by President Luiz Inácio Lula da Silva, but in the end, that did not happen. From that point of view, Rio was perhaps less grand. Remember that in New Delhi, we had Prime Minister Rajiv Gandhi, in Kyoto the Emperor of Japan, and in Beijing the (then) Vice-President of the People's Republic, Xi Jinping. Still, the Rio meeting was fine—there was, of course, no Pluto and the financial issue had been understood and the recovery efforts appreciated! The success of IYA illuminated the meeting, we introduced the Strategic Plan, which was very well received and so, despite a few safety issues relating to the place itself, I believe that the General Assembly in Rio went well.

After Rio, we added a distinct IAU outreach activity to the original Strategic Plan. I had long thought about it, and when Norio Kaifu offered to host the activity in the shape of the IAU Office for Astronomy Outreach, everything fell into place.

As regards the internal matters of the IAU, we began the effort to strengthen the Divisions. Thus one of the yearly meetings also involved the Division Presidents that brought science to the centre of our discussions in the Executive Committee, and they were given more and more responsibilities. The question of the IAU division structure, begun by Lodewijk Woltjer, has remained with us, with the next step taken by Thierry Montmerle under Norio Kaifu's presidency. In this sense, I am glad to see that the IAU has managed to continually regenerate itself.'

Cesarsky was the first (and only) female Director of basic research at CEA, the first (and only) female Director General of ESO, the first (and only) female High Commissioner for Atomic Energy in France, and the first female President of the IAU; but there, she is happy to see distinguished successors!

'I'm usually not in favour of highlighting this fact. But the very good thing about it is that at the IAU, there will have been two women after me. As regards the gender issue in general, I believe that in astronomy the situation is less bad than in many other disciplines. And in any case, the problem is linked to the culture and the traditions in the different countries rather than to our science. In fact, I've recently chaired the HST Time Allocation Committee and there they've essentially achieved parity. It was an absolute pleasure to run this committee. A bit earlier, I attended an ESA meeting in Venice regarding the JWST, and once again it had excellent participation by women.'

(Continued)

After her term as IAU President, Cesarsky continued as French High Commissioner for Atomic Energy until 2012 and held or holds many functions since, e.g. for CERN, for ESA, for the French Ministry of research, for the French Academy of sciences, for the European Commission, and for various ongoing astrophysics projects. In the course of her long career, she has received numerous awards, honorary degrees and memberships, including the National Academy of Sciences (USA), the Royal Swedish Academy of Sciences, and the Royal Society of London. She is Grand Officier de l'Ordre National du Mérite and Commandeur de l'Ordre la Légion d'Honneur.

The International Year of Astronomy in 2009

Already before the Pluto affair, the IAU leadership had realised the enormous popular appeal of astronomy and decided that the Union should get involved more actively in outreach projects. In fact, preparations for a major outreach event had already started. If anything, the 'Pluto affair' provided extra motivation to make it a success.

In 2002 IAU President Franco Pacini saw an opportunity to stimulate astronomy outreach worldwide: in 2009, it would be 400 years ago since Galileo first turned a telescope to the sky and made an astonishing series of discoveries that changed astronomy forever. Would it not be an excellent occasion to declare an International Year of Astronomy, with many national and international activities promoting the most beautiful of sciences? The International Year of Physics in 2005, 100 years after Einstein's *annus mirabilis*, provided an example. Pacini formally proposed this idea to the IAU General Assembly in Sydney in 2003, in the form of a resolution calling for 'the year 2009, the 400th anniversary of Galileo's accomplishments and the real birth of modern telescopic astronomy, [to] be declared the "International Year of Astronomy", in which the potential [of] astronomy to enlighten and enrich humans will be brought to the largest possible audience all over the world.'

With the approval of the General Assembly, the Italian astronomers brought forward the proposal to the Italian government that in turn introduced it to UNESCO for endorsement by the 2005 UNESCO General Conference. In parallel, a dedicated IAU Working Group, chaired by Franco Pacini was established.

Meanwhile, at UNESCO, the proposal was well received and endorsed through a Proclamation, but it was also pointed out that 'International Years may be proclaimed by the United Nations only during the annual General Assembly meetings in autumn at the request of one (or more) Member States, one year in advance.' On this basis, the Italian Government declared itself ready to submit the proposal to the UN General Assembly in New York in the autumn of 2006. Buoyed by the initial success, the IAU thus began preparing for an ambitious event. At the 2006 IAU General Assembly in Prague, the topic was clearly in the air and the President-Elect, Catherine Cesarsky (also Director-General of ESO), began to assemble a team to be charged with developing and implementing what was becoming known as 'the 2009 IYA'.

Already in 2006, the first elements of the IYA2009 were ready. Shown here is a presentation press conference at the Prague GA regarding the 'Universe Awareness' (UNAWE) project. Seen from left to right: George Miley (Chairman of UNAWE), Carolina Ödman (UNAWE Coordinator at the time), Cecilia Scorza (astronomy education expert) and Claus Madsen (UNAWE Co-Chairman). Photo: ESO Archive

For the IAU, it had been smooth sailing so far, but this was not to continue. Just weeks after the General Assembly, it appeared that the Italian UN Mission had encountered massive political opposition to the idea. This was not because of the subject matter itself, but rather due to general political skirmishes, as they are often fought out at the only forum where all the countries of the world will meet. Right at this time, especially the Western countries were increasingly reluctant to support any new proposals for 'International Years of…', of which they thought there were too many and often of dubious value. Hence, the Italian mission concluded that it would not be possible to gather enough support in the General Assembly, and rather than suffering a humiliating diplomatic defeat, refrained from presenting the proposal.

At the IAU, however, technical preparations continued at an ever-increasing pace. A dedicated working group of the Executive Committee was established, chaired by Cesarsky, and a secretariat was set up at ESO, headed by Lars Lindberg Christensen and with Pedro Russo as 'IYA Coordinator'. In parallel, with substantial support from ESO, the IAU mounted a major 'behind-the-scenes' diplomatic effort towards the UN member states, spearheaded by Claus Madsen and Enikö Patkós. Luckily, by the early autumn of 2007, the Italian government had gained enough confidence in the effort to finally submit the proposal to the UN in New York. This was backed up by the visit on 1 November of an official IAU delegation to the UN Headquarters who argued their case in a crowded conference room.

The official IAU delegation outside the United Nations Headquarters in 2007. From left to right: Claus Madsen (ESO), Beatriz Barbuy (IAU Vice President), Robert Williams (IAU President Elect), Catherine Cesarsky (IAU President), Luigi de Chiara (Italian UN Mission), Kevin Govender (SALT), Nella Le Moli (INAF) and Leopoldo Benacchio (INAF). Photo: ESO Archive/C. Madsen

Subsequently, Italy submitted the official proposal to the General Assembly's Second Committee. The proposal was now supported by 30 countries. There were some small, but remarkable diplomatic scoops: among the co-sponsors were Algeria and Israel, as well as Armenia, Turkey and Greece—countries that rarely see eye to eye on any subject. In the end, the Second Committee delivered a consensus recommendation in favour of the proposal to the General Assembly, and in the afternoon of December 19th, 2007, (New York time) the proposal was finally approved by the Plenary Assembly.

The UN resolution designated UNESCO as the lead agency for 'The Year', but explicitly named the IAU and ESO and 'astronomical societies and groups throughout the world' as actors. While UNESCO would be the lead organisation in the frame of the 'UN system', the worldwide implementation would, therefore, be a collaborative effort involving everyone with an interest in astronomy. Indeed, it was now up to the IAU and its partners to deliver. And together, they did, with astounding success. With the diplomatic challenges resolved, the focus shifted to filling the IYA with content, to raise the necessary funds, and to secure participation by establishing national nodes across the globe. Already in March 2007, i.e. before the final approval by the UN, a meeting of national node representatives—also known as National Single Points of Contacts or 'SPoCs'—took place at ESO to ensure that no time would be lost on preparing the activities during the Year.

THE UNIVERSE
YOURS TO DISCOVER

INTERNATIONAL YEAR OF
ASTRONOMY
2009

The official logo for the International Year of Astronomy 2009. Photo: IAU Archives

Some discussion occurred over the logo, perhaps not surprisingly. In the end, a consensus was achieved over a graphical representation of an adult and a child hand in hand looking up towards the sky. It was based on an original design by Edmund Janssen (ESO), but modified to make it gender and race 'neutral'. It also came with a slogan for The International Year: 'The Universe—Yours to Discover', underlining that the IYA was indeed intended for every citizen on Planet Earth. This was a short-hand articulation of the vision for the IYA, elaborated by the IAU, 'to help people rediscover their place in the Universe through the sky, and thereby engage a personal sense of wonder and discovery'.

In terms of content, the core elements were the so-called 'cornerstone' projects—activities thought to be of high visibility and with a global reach. They would be complemented by what eventually grew to become a huge number of events and activities organised at the national level, under the responsibility of the SPoCs or by the IYA institutional partners. The cornerstone projects would receive seed funding from the IAU, and would all follow a common identity and be subject to central coordination. The cornerstone projects were:

1. 100 Hours of Astronomy: an event 400 years in the making
2. Galileoscope: millions looking at the sky
3. Cosmic Diary: the life of an astronomer
4. The Portal to the Universe: a one-stop universe of news
5. She is an Astronomer: breaking down misconceptions
6. Dark Skies Awareness: seeing in the dark
7. Astronomy & World Heritage: universal treasures
8. Galileo Teachers Training Program: bringing astronomy into the classroom
9. Universe Awareness: one place in the Universe
10. From Earth to the Universe: millions looking at the sky
11. Developing Astronomy Globally: astronomy for all
12. Galilean Nights: The Galileo Experience

In April 2009, the *'100 Hours of Astronomy'* cornerstone project 'took off' with a 24-hour webcast called 'Around the World in 80 Telescopes' from astronomical sites across the world. This was characterised as a 'star-party', with a second one in October—the 'Galilean Nights', designed for members of the public to observe the sky themselves. The 'Galilean Nights' saw about 1 million people enjoying 'street astronomy', with amateur astronomers and professionals taking to the streets with telescopes, offering a view of interesting celestial objects to the public.

For those who might not have the chance to participate in a live observation, the cornerstone project *'From Earth to the Universe'* presented itself as an opportunity to study the heavens through a series of large photographs of some of the most spectacular astronomical views. The collection of pictures was put on display in about 1000 locations in 70 countries, in public places such as shopping malls, parks, train stations, hospitals, libraries, etc.

Aficionados who would frequent the internet, perhaps on a daily basis, to stay abreast of the news from the world's observatories were provided with a 'one-stop shop' in the shape of the central website *Portal to the Universe*, another cornerstone project.

In the *Cosmic Diary* cornerstone project, 60 professional astronomers blogged about their lives, work, hobbies and interests, producing thousands of entries, while *She is an astronomer*, through a website, but also with meetings and media activities, sought to raise awareness and recognition of female scientists. Like the other cornerstones, this was indeed a global activity. Examples also include astronomy lectures in Saudi Arabia especially for women, and a meeting in Tunisia in connection with the International Women's Day, with lectures by women scientists.

In addition to the cornerstone projects, the IYA generated a host of activities, such as the special Galileo Mobile Project that brought astronomy to young children in Chile, Bolivia and Peru, travelling 5000 kilometres through the Andes. The project travelled with a van fully equipped to offer observing opportunities, including the IYA's handy Galileoscopes that were subsequently donated to the schools. The project finished with a star party in the Chilean coastal town of Taltal. Photo: ESO Archive

One of the most auspicious projects, *Dark Skies Awareness* served to draw the attention of the general public (as well as decision makers) to the ever-growing problem of light pollution. Activities included events such as 'Earth Hour', 'World Night in Defence of the Starlight', 'Dark Skies Discovery Sites' and 'Nights in the National Parks'. This contributed to elevating the issue to the political level and led to policy actions in several countries. The vision of 'rediscovery' was also the thread of the UNESCO-led cornerstone project *Astronomy and World Heritage*, demonstrating 'the importance of astronomical heritage in terms of the enrichment of the history of humanity, the promotion of cultural diversity, and the development of international exchange.

Obviously, Galileo played a significant role in the IYA, albeit mostly by giving his name to some of the cornerstone activities: One was the *'Galileoscope'* project, producing and delivering cheap telescopes (essentially modern-day versions of Galileo's telescope) to individuals, schools and 'organisations'. An impressive 175,000 telescopes were distributed during the year, some sold, others donated. A second one was the *'Galileo Teachers Training Program'*, introducing some 5000 teachers worldwide to astronomy and astronomy teaching.

Finally, two other cornerstone projects focussed on the development of astronomy and on astronomy for development, essentially laying the *seeds for what would later become main parts of the IAU Strategic Plan 2010–2020*: *'Universe Awareness'* (also known under its shorter name 'UNAWE'), specifically targeting young, often disadvantaged children in the developing world or from minority groups in derelict inner cities in the West to the inspirational aspects of modern astronomy, and *'Developing Astronomy Globally'* to foster astronomy teaching at all levels.

Beyond the cornerstone projects, several movies, both documentary and fiction, were produced. They included 'Eyes on the Skies', which received several prestigious film awards, 'BLAST!', a fantasy film screened by the BBC, and other films graced with numerous awards.

Apart from the centrally coordinated projects, there were many national and local activities as well: observing opportunities, exhibitions, school lessons, television shows, and many more. Amateur astronomer societies naturally played an important role. This was one of the keys to the success of the IYA: it activated local committees, and provided a global framework to advertise their activities. In the end, there were 148 national 'nodes'—many more than IAU national members.

The IYA was officially opened at a grand ceremony at the UNESCO Headquarters in Paris in January 2009. Photo: IAU Archives/J.F. Delgado

The International Year of Astronomy was launched on 15/16 January with a spectacular opening ceremony at the UNESCO Headquarters in Paris. Fittingly, it came to an end with a formal closing session in the magnificent Aula Magna of the University of Padova on 9/10 January 2010, attended among others by Franco Pacini, by then clearly marked by a rapidly deteriorating health. The title of the closing session was *'Astronomy Beyond 2009'*, signalling that the IAU and the worldwide astronomical community had every intention of capitalising on the huge effort invested in The Year. Indeed, with a staggering 148 countries, 73 organisational nodes and partners and 22 media partners to deliver a plethora of exciting, high profile events, the IYA had something to offer for everybody. The 1433-page strong final report of the IYA estimated that 815 million people were exposed to astronomy during the year, which would make it the largest outreach project ever.[11]

If anything, the report contained not only a treasure of hard-won experience from public activities, but also pointed to such that could—indeed ought to—be carried on after the closure of the IYA itself. For this purpose, the IAU adopted the term IYA legacy activities. But the impact of the IYA went beyond the strongly

[11] Russo and Christensen (2010).

increased visibility of astronomy. Keeping in mind the considerable political resistance in the UN regarding international years the IYA, by its very success, created the path to new international years, such as the *International Year of Light*.

George K. Miley: Outspoken Science Diplomat Arguing the Social Responsibility of Astronomy

Vice-President (2006–2009), Vice-President (2009–2012)
  Interview conducted in Leiden on 20 April 2017
  His solid academic and scientific contributions notwithstanding, it is hard not first and foremost to describe George Miley as a visionary idealist. As IAU Vice President (2006–2012) he both finessed and broadened his ideas about the role the IAU could—and in his view ought to—play in the field of education and public awareness of science.

George Miley at his home institute in Leiden. Photo: C. Madsen

'As regards astronomy, my first inspiration was when my father read to me "The War of the Worlds" by H.G. Wells. I subsequently wrote to Hermann Brück at Dunsink Observatory near Dublin—I must have been 8 or 9 years old—and he invited me to come and visit the place. I later studied physics at the University College Dublin. It was at the time when radio astronomy was becoming very exciting, so I went to do a PhD at Jodrell Bank in 1968. After a spell as Post-Doc at the NRAO I came to Leiden, where Jan Oort offered me a job to work with the newly established Westerbork Radio Observatory.'

Miley relinquished his position as Director of the Leiden Observatory in 2003, but he was still full of energy and ideas. So when he was awarded the Royal Netherlands Academy of Arts and Sciences Professorship, it gave him the opportunity to think about—and develop—the Universe Awareness initiative. This was an activity specifically targeting underprivileged children across the world—from remote rural areas in developing countries to derelict inner cities in parts of the industrialised world. One aim was to expose them to the beauty of the Universe and the power of science in interrogating and understanding the natural world

(Continued)

around us. A second equally important goal was to use the enormity of the Universe to stimulate a sense of world citizenship at an age when a child's value system was being formed.

'I was educated by Jesuits whose motto was attributed to be "Give me a child until he is seven and I have them for life". I've seen what can be achieved when working even with very young children. With Universe Awareness we wanted to reach some of the children that were among the most vulnerable when it comes to being absorbed by hopelessness or perhaps even falling prey to sinister ideologies and beliefs.

I've been to several IAU General Assemblies, starting with Sydney in 1973, but I wasn't really involved with the IAU before 2006 and the Prague GA. Just before that, I was phoned up by Ron Ekers, the IAU President, asking me if I wanted to become an IAU Vice President. I suspect that the reason had to do with the Universe Awareness Initiative, which I had started two years earlier. That seemed to be proceeding well and we had endorsements of the programme from the IAU and, in fact, from Ron Ekers himself.

I didn't come in with clear plans for my term at the IAU, but somehow, it was clear that since we had started Universe Awareness, this was a direction I wanted to take. In fact, we had a launch event for Universe Awareness at the Prague meeting.

In early 2007, the Executive Committee met at the SAAO in Cape Town. There was a suggestion for developing strategic plans for the various activities of the IAU and I volunteered to lead the work in developing a plan to use astronomy to stimulate international development. This plan would build on education and outreach initiatives within and outside the IAU and on the forthcoming International Year of Astronomy. Indeed the Cape Town meeting provided a strong stimulus to go ahead with this strategic plan. Working with Karel van der Hucht and Ian Corbett, this was followed up with a stakeholders meeting that I organised in Paris. There were a dozen people, including members of Commission 46 on Education and Development. After that, I wrote the first draft document and circulated it. Johannes Andersen had done an excellent job in reorganising Commission 46. The Commission comprised some good people, but for several years there had been little new blood or new initiatives. Perhaps more important, the Commission was playing no significant role in the forthcoming hugely important International Year of Astronomy (IYA). This was difficult for me to understand.

In any event, given the IYA and its potential for creating a lasting legacy, I felt that there was room for the IAU to do something bigger as an IYA follow-up. Back then, Kevin Govender was leading an IYA Cornerstone Project on using astronomy for development. I was very impressed by that—young people, lots of enthusiasm and so on. Other encounters and visits, among them one to Mongolia, strengthened my conviction that the development theme was the right one and one with great potential. People like Ian Corbett and Bob Williams, then IAU President Elect, gave strong support to the ideas contained in the draft Strategic Plan. Ed Guinan, himself a member of Commission 46, also supported the work with great enthusiasm, as did Karel van der Hucht. The Strategic Plan adopted the same rationale as that used by the new South Africa for supporting astronomy—namely that because of its inspirational aspects, it can play a unique role in building technological, scientific and cultural capacity.

(Continued)

We were extremely happy to see the Strategic Plan endorsed by the Rio General Assembly and we quickly launched a Call for Proposals for hosting the new Office of Astronomy for Development (OAD). This was a crucial element of the Plan, needed to coordinate its implementation. About 30 letters of intent and 15 concrete proposals for hosting the office came in. Among them was the proposal from South Africa. It was at the time when South Africa was bidding also to host the SKA "megafacility", so politically, it made great sense. However, it was also a time when the astronomical community there was facing problems of leadership that could easily have resulted in no South African proposal being submitted. However, Kevin Govender and Patricia Whitelock together with several people inside SAAO and the National Research Foundation of South Africa managed to submit their proposal, which was unanimously chosen by the IAU Executive Committee as the winning bid. After widely advertising and conducting a global search, Kevin Govender was appointed as Director of the OAD. Since I was a personal friend of Kevin and some other applicants, I recused myself from the hiring process, but I was delighted in the outcome. It was a perfect choice. In 2015 a high-powered visiting committee reviewed the OAD, not only finding its performance outstanding and strongly recommending an extension until 2021, but even appealing to the IAU to continue the activity beyond the 10-year running time of the Strategic Plan. And in 2016, Kevin Govender and the IAU were jointly awarded the Edinburgh medal for the IAU Astronomy for Development programme.

From the launch of the IAU Global Office of Astronomy for Development at the headquarters of the SAAO: OAD Director Kevin Govender and Ms. Grace Naledi Pandor, South African Minister for Science and Technology. Photo: IAU/OAD Archives

(Continued)

Universe Awareness, the OAD and the IAU Strategic Plan did not come about by coincidence. Looking back, the period during which we developed the Strategic Plan and started the OAD was highly auspicious. We had a President and a General Secretary who strongly supported the initiative—and we had circumstances in South Africa that made them interested in contributing to it. As so often, success is dependent on having the right people in the right place and at the right time—and we had a number of really special people with remarkable talents and charisma. There was Kevin Govender, but also people like Khotso Mokhele, the impressive past CEO of the South African National Research Foundation, who became involved with the OAD as Vice Chair and later Chair of the OAD Steering Committee.

Since its establishment, the OAD has organised activities across the globe. Seen here are the so-called 'Gaza Ambassadors of Mars'. Photo: IAU/OAD Archives

I've been privileged to work as a scientific researcher for all my professional life. And yet, I've always felt an obligation to engage with society at large and give back to society what I could. I guess this partly goes back to the influence of my father, who was a lawyer—and an armchair socialist—and partly to what I saw in Ireland during my youth, a society in desperate need for change.'

Miley has always been politically aware and he has a keen understanding of the cultural dimension of astronomy and its ability to act as a bridge between people—and perhaps even peoples.

'I went to Mongolia in connection with a TAD School—Teaching Astronomy for Development—around 2007 or 2008. During my visit I met two astronomers from the Democratic Peoples Republic of Korea (DPRK). They were delightful people with a good sense of humour. Getting to know them was an eye-opener, because in the Western World we have these stereotypes about North Koreans. I took the opportunity to suggest that North Korea might consider rejoining the IAU. They appeared to be interested and I had hoped to see them at the Rio General Assembly. Unfortunately, they did not come to Rio, so before the Beijing General Assembly I made a renewed contact with North Korea, writing to their

(Continued)

Academy of Science in Pyongyang. And lo and behold, it worked, as the North Koreans did attend the Beijing meeting. I know that as IAU Vice President I stuck my head out, but I don't have any regrets. Later, two North Korean astronomers came to Leiden for half a year and I've been invited to pay a visit to the DPRK. In my view the perspective given us by the enormity of the Universe gives astronomy a unique role in engaging in science diplomacy. If through the IAU we astronomers can have a way to communicate across—and despite—the political boundaries, then we've done something really important for the world. What the IAU has given over and beyond the normal science conferences has been the General Assemblies, where astronomers from across the world and working in very different fields can meet. The Regional Meetings have played an increasingly important role in bringing astronomers from developing countries together and helping to break their relative isolation. In my view this is one of the most important present functions of the IAU'

Robert Williams: A New Epoch for the IAU

Vice-President (2000–2006), President-Elect (2006–2009), President (2009–2012)
    Interview conducted in Trondheim on 20 June 2017
    'For me, if there were an overall description of the historical activity of the IAU it would be 80 years of being internally focussed, appropriately so for those times, to one that in the past two decades has evolved into a more outwardly looking organisation.'

Bob Williams at the Starmus IV festival in Trondheim in June 2017. Photo: C. Madsen

'I was the first of our large extended family to attend university. As a 12-year old in Southern California, I was required to take a semester course of general

(Continued)

science, and I found it fascinating. We were given booklets on a selection of sciences and I would read ahead. I began reading the booklet about astronomy. There was a section about the planet Mars that showed a picture of the red planet taken with the Yerkes 40" refractor and it blew my mind. I had no concept of other planets, and Mars even had some similarities to earth. I was possessed! That day I became an astronomer. After school, I walked home and from my mother's sewing kit I grabbed a small magnifying glass that she used when threading small needles. I figured that I might see the canals on Mars that Schiaparelli and Lowell believed existed by magnifying the picture in my booklet but, of course, all I resolved were a bunch of little dots. No matter, I was hooked.

There's not been a day in my life ever since that time when astronomy has not been one of the driving forces in my life. I obtained my PhD in 1965 at the University of Wisconsin with Don Osterbrock as my thesis advisor and was offered a job at the University of Arizona's Steward Observatory even before I finished my thesis—at 24 years of age I was an Assistant Professor in Tucson teaching the physics of astrophysics.

At that time Bart Bok was director of Steward and chair of the department. Bart was a strong advocate of the IAU and he encouraged all faculty members to become active in the IAU. He himself became an IAU Vice President a few years later. I was interested in international aspects of astronomy and the IAU was the international organisation for astronomy, so I applied for membership. I first attended the 1967 General Assembly in Prague and its associated Symposium 34 on planetary nebulae. For quite some time my primary links with the IAU amounted to participation in symposia. The IAU was internally focused at that time, meaning that it attended primarily to the regulation of international astronomy at the professional level. As a young astronomer the large generational gap between IAU leadership and myself caused me to feel understandably distant from IAU "machinery", but even in that context I acknowledged that the IAU did seem to serve a useful purpose in regulating various aspects of astronomy in a beneficial way.'

In 1983 Williams decided to look for new professional challenges and after several short-term research appointments at ESO/Garching and NASA/Ames Research Center he was offered a research and administrative position at Cerro Tololo Interamerican Observatory (CTIO), where he remained until 1993.

'In the Southern hemisphere, with a different culture and significant directorial responsibilities, those years were magical for my wife Elaine and I. Astronomy in Chile was captivating. After eight years we decided to return the U.S., and I was fortunate to be offered the Directorship of the Space Telescope Science Institute (STScI) in Baltimore.

I joined STScI at the time of the First Servicing Mission that corrected the Hubble Space Telescope's spherical aberration. It became very clear that outreach efforts for HST needed a boost because of the poor reputation in the public of a very expensive HST that was flawed. One of my first tasks at STScI was to lead development of a "strategic plan" for the Institute. That effort defined a major strategy for the Institute, in which public education and outreach featured prominently.

As for working with HST, could one hope for more good fortune than being connected to a science programme that grips the attention of the scientific world? But time passes on and I wished to have the end game of my professional career

(Continued)

devoted to research and lecturing rather than spending 95% of my time on administrative and managerial duties. So, in 1998 I stood down as Director of the Institute in order to return full time to research and graduate teaching. I admitted nevertheless to still being comfortable in taking on some managerial oversight roles and setting science directions, and so it was that supporting the IAU in its functions seemed to me to be something that might be of value and that I might consider.

From the EC68 meeting at the STSci in Baltimore in early 1996: (From left to right) Franco Pacini, Bob Williams (host), Bob Kraft and Virginia Trimble. Photo: J. Andersen

About this time I had a chance conversation with Ron Ekers in which I expressed interest and support for the directions the IAU might take. Some months later, I received an email from the IAU Executive Committee asking me if I might be willing to be considered for election to the EC. I responded positively and at the year 2000 General Assembly in Manchester was elected as one of the three IAU Vice Presidents for the next triennium.'

Thus began 15 years of close association with the IAU consisting of two periods as Vice President, a term as President-Elect, then as President and finally a three-year term as past-President in an advisory capacity to the Executive Committee. During his tenure, early on he saw the need for revising the IAU statutes, by-laws, and working rules (S/B-L/WRs).

'The lead in this effort was Johannes Andersen, who had recently completed his term as General Secretary of the Union and was very familiar with these formal documents that governed the policies and procedures of the IAU. Johannes and I saw our task in the same light: the three documents were a patchwork of modifications made over the years and their entire style and verbiage was archaic and often contradictory. We agreed; we had to rewrite the documents and bring them into the modern world. Johannes did the lion's share of work in rewriting

(Continued)

and simplifying the texts, correcting shortcomings and inconsistencies, suggesting new procedures, and so on. The resulting S/B-L/WRs were put before the National Members, who suggested further revisions before their final approval in 2003.'

The adoption of the documents turned out to be a crucial component in changing the culture, operation, and outlook of the IAU. One important feature that was reinstated into the revised S/B-L/WRs—albeit first at the 2006 GA—was that at General Assemblies, individual members of the Union would henceforth vote upon resolutions of a scientific nature, not the national members as had previously been the case. After the famous debate and vote, by a small fraction of total Union members present, on the status of Pluto at the 2006 Prague General Assembly even this revised procedure would soon be changed in a major way for the future: on-line voting for the entire IAU membership.

The Rio General Assembly in 2009 took place during the International Year of Astronomy (IYA), and in many ways the IYA effort led to the realisation that the IAU truly needed a strategic plan.

'Catherine Cesarsky led the effort towards the IYA brilliantly. It represented by far the most extensive planning ever undertaken between the IAU executive and the national members. The IYA was a huge worldwide success in connecting astronomers in all nations to the public. My involvement was marginal compared to Catherine's. I helped, of course, by taking part in the IAU Delegation visit to the UN Headquarters in New York. But much of the preparations took place in Europe, and correctly so. Firstly, the proposal came from Italy; secondly, ESO was ready to play a strong role; and also because of the links with UNESCO as the leading UN Agency in this matter, it felt natural indeed that the Europeans would take the lead in this.

The IYA touched upon an area of common interest that many of us on the EC shared: education and public outreach. The motivation for creating a strategic plan emerged from the planning of the IYA. Vice President George Miley became involved and drove the effort forward, not the least building on his early experience with the Universe Awareness—or UNAWE—programme. This led us to propose creation of the Office of Astronomy for Development (OAD) that eventually was approved and created in collaboration with the South African government, headed by Kevin Govender and located in Cape Town. The OAD was a truly seminal undertaking by the IAU. For the first time, the Union provided significant operational funding for a major unit whose efforts were directed outside the IAU to the public.

(Continued)

The IYA GA in Rio included a 'town Hall meeting' to discuss the Strategic Plan. Seen here are incoming IAU President Bob Williams and IAU Vice President George Miley. Photo: IAU Archives

For me, my own experience with the STScI strategic plan and the appreciation of the importance of public outreach came in handily and I was really happy to contribute to its development with George's leadership.'

Williams' term as President ended with the Beijing General Assembly in 2012. This was only the third General Assembly in Asia, following the ones in Delhi and Kyoto.

'We worked very well with the Chinese. Our cultures are different, but the Chinese delegation was always responsive to suggestions and we developed a great relationship. We were, of course, keen to hold the Assembly at the new international conference centre that was being built in connection with the Olympic games. This did work out and our dealings with the Chinese, especially Gang Zhao and Yanchun Liang, were wonderfully smooth. The current President of the People's Republic, at the time Vice President Xi Jinping, even picked up my reading glasses when they fell from my pocket during the opening ceremony. There's a crack in the frames, but I still use those glasses.'

(Continued)

Next to the high-level participation, the Opening Ceremony of the IAU GA in Beijing offered a spectacular artistic programme. Photo: C. Madsen

One of the major developments during the last two decades has been the growing cooperation with philanthropic foundations, the Gruber and the Kavli Foundations in the awarding of international prizes.

'It began in 2000 when the Gruber Foundation approached the IAU about its Cosmology Prize in 1998. The Grubers realised the value of having the imprimatur of international scientific organisations in their prizes. The IAU agreed to have the Executive Committee recommend members of the Cosmology Prize selection committee in return for the Foundation creating funding for a Gruber postdoctoral fellowship in astronomy that the IAU would manage. This relationship has been very fruitful for both organisations.

Subsequently, the Kavli Foundation contacted the IAU via the Norwegian Academy of Science and Letters, asking for IAU collaboration in recommending members of the selection committee for its Kavli Prize in Astrophysics. As with our association with the Gruber Cosmology Prize, this relationship has been beneficial to the IAU through the funding of the IAU International Schools for Young Astronomers.

(Continued)

Astronomy is a magnet for young people and can become a gateway to the sciences for the next generation. With the OAD and OAO the IAU can play an important role in supporting outreach activities across the world. The picture here is from the "open day" at the IAU GA in Honolulu. Photo: C. Madsen

Looking back on my tenure on the Executive Committee I find great satisfaction in having been part of a creative group of astronomers who realised that astronomical research, both ground and space-based, was impacting the public in an exciting way and that we should change the face of the IAU to be a more outward looking organisation. Astronomy has given us new insights into the universe and humanity itself, and the IAU has helped bring these to the attention of the world. Whether one sees the work of astronomers on television or reads about it in Die Welt, Le Monde, China Daily, or the New York Times it is clear that astronomical research is not simply fascinating, but also important in understanding life and our relationship to the cosmos. The relationship between the public, astronomers, and the IAU has been a very good symbiosis. It has happened in our lifetime and it feels great to have been a part of it!'

Norio Kaifu: IAU and the Rise of East Asian Astronomy

Vice-President (1997–2003), President-Elect (2009–2012), President (2012–2015)
    Interview conducted in Tokyo on 27 May 2017
    Norio Kaifu has not only witnessed the evolution of Japanese and East Asian astronomy from the dire early post-war period until today, he has also been an

(Continued)

important player, helping to foster and shape its impressive rise, not the least when it comes to international cooperation. His long-standing involvement with the IAU has been important in this context. But it also influenced the Union itself.

Norio Kaifu at the Mitaka Headquarters of the National Observatory of Japan. Photo: C. Madsen

'I've spent a substantial part of my young life on large telescope projects, including the Nobeyama Radio Observatory (NRO) with its 45-metre Millimetre-Wave Telescope. That was a big challenge for Japanese astronomy and at the time, it took my full attention, but in a way, it also made me aware of the need for international cooperation.

The telescope was completed in 1982 and we decided that the XVIIIth IAU General Assembly in Patras in the same year would provide a good opportunity to show first results. The Patras meeting thus became "my first" IAU General assembly.'

Born in Niigata in September 1943, Norio Kaifu entered into University of Tokyo in 1962. His initial scientific focus was the study of the interstellar molecule in mm-waves, moving on to star-formation and more recently to exo-planet studies.

'I was really interested the question of life in the Universe. This interest was definitely triggered by the 1968 discoveries of ammonium and water-vapour molecules, and the Formaldehyde detection in 1969. I was excited. These are organic molecules in interstellar space! However, in Japan, we didn't have any radio telescope at the time. In fact, I based my own PhD thesis on an expanding molecular ring in the Galactic Centre on published data from the Parkes 64-m Telescope in Australia. Construction of the first radio telescope in Japan, a 6-metre millimetre telescope started in 1968, and led to the first detection of para-formaldehyde molecules in 1972, the very same year as I completed my PhD.

(Continued)

We organised the IAU Symposium 115 on Star Forming Regions in 1985 in Tokyo. In 1997, at the time of the XXIIIrd General Assembly in Kyoto, I became IAU Vice President. I found the international aspect very attractive. In those days, the workload on the Vice Presidents was light, but it nevertheless gave me a taste of the "inner life" of the IAU.

My term ended with the General Assembly in Sydney in 2003. But then, in 2006 I became involved in the preparation of the 2009 International Year of Astronomy (IYA 2009), when Catherine Cesarsky, then IAU President, asked me to join the IYA 2009 working group. I have always taken much interest in public outreach and, in fact, written more than 20 books for the general public. There is much activity in Japan in this field with no less than 300 "Open Observatories" as well as 300 planetaria operated by local governments in the country. For the IYA, we obviously also involved the amateur astronomers, creating a common framework for all astronomical communities of Japan. I enjoyed the IYA 2009 work tremendously. By then, I had become very interested in the IAU, not the least because of the IYA, and I thought that the IAU had an important role to play in astronomy outreach and education. And so, when Sadanori Okamura asked me if he could suggest my name for the task of IAU President Elect, I accepted the challenge.

Members of the IYA Executive Committee working group, seen here at the GA in Rio. From left to right: Ian Robson, Norio Kaifu, Kevin Govender, Mariana Barrosa, Claus Madsen, Catherine Cesarsky, Dennis Crabtree, Mary-Kay Hemenway, Lars Lindberg Christensen and Pedro Russo. Photo: IAU Archives

(Continued)

After the IYA2009, we managed to establish the Office of Astronomy for Development (OAD), thanks to strong support from the South African authorities, and later the IAU Office for Astronomy Outreach (OAO) in Tokyo, with equally strong support by the NAOJ. While the OAD is about "development of astronomy" itself, the OAO has a different mission, e.g. keep and strengthen the network of professional and amateur astronomers constructed in the course of the IYA, and communicate with the worldwide public by organising a series of outreach activities etc. For that purpose, the OAO is working hard to expand its network of national contact points. Also, the initiative of opening up for the naming of exo-planets to the public is an important action. Thus the OAO and the OAD complement each other well.

Together with Bob Williams, who preceded me as IAU President, it became my task to support the OAD and to establish the OAO with the strong support by the two General Secretaries, first Ian Corbett and subsequently Thierry Montmerle.'

The establishment of the OAD in Cape Town and of the OAO in Tokyo is both indicative of the growing importance of these regions in IAU, and helping to boost the IAU's global activities even further and wider.

Kaifu's term as President started in 2012 at the XXVIII General Assembly in Beijing and ended with the General Assembly in Honolulu three years later.

'This was a happy coincidence for me because China is our neighbour and I am a strong supporter of East Asian regional cooperation in astronomy, and Hawai'i used to be my home place when I was in charge of the Subaru Telescope Construction.'

Kaifu's strong involvement in major research infrastructure projects and also in science outreach notwithstanding, he leaves an important legacy in the area of regional cooperation in astronomy.

'I was in charge of the construction of Nobeyama 45-metre telescope known for its extremely high accuracy as a mm-wave telescope. It had become a world-class facility, worked very well and produced good science. And yet, scientifically, I felt a "bit lonely". We didn't know about astronomy in China and the Chinese astronomers didn't know about us. We cooperated with astronomers in Europe and in the USA. The Chinese and Koreans did so, too. But we did not cooperate with each other, and I felt very strongly that this ought to change. We're neighbours living close together, with a common cultural background and way of thinking. To change this, I organised bi-annual symposia of East Asian Meeting of Astronomy (EAMA) with colleagues in China and South Korea. This idea was also discussed informally in the Asian-Pacific Regional IAU Meeting (APRIM) in Pune 1993.

While the Regional Meeting was—and still is—important for scientific exchange, I wanted to move to a different level—one of a closer regional cooperation especially in observations and instrumental developments. This took us on a particular path. In particular, I worked with Cai-Ping Liu, who hailed from Taiwan, had received her astronomy education in Japan, but then moved to Mainland China. She is a good friend of me, and she invited me to visit the Purple Mountain Observatory in 1978, just after the Cultural Revolution. In China, she also became a member of the People's Congress, and thus quite influential, and as a professor of the Purple Mountain Observatory, she supported sending astronomy students to Japan. We started EAMA in 1990 with the first

(Continued)

China-Japan workshop "Star Forming Regions" in Huangshan in China. Another important person was Se-Hyung Cho of Korea, who joined the Huangshan workshop and invited the second EAMA meeting in Daejeon in Korea. That was, in fact, the first international astronomy meeting in Korea. The EAMA was appreciated by Asian astronomers broadly, but even so, I will not hide that the idea of regional cooperation in Asia was not necessarily popular among all astronomers, as the cooperation with western world was regarded as "more efficient", so it was an uphill battle.

Our idea was to first create a grass-roots initiative focussing on practical programme, such as the East Asian VLBI Network and East Asia Young Astronomers Meeting, of which we are very proud as results of the EAMA initiative. Common activities also include the development of instrumentation, joint observations and site searches for future telescopes in Asia.

But EAMA had no funds of its own, so we needed to enlist institutional support. In 1990, when we started EAMA, Japan was the only East Asian country that had established a national astronomical organisation, the National Astronomical Observatory of Japan (NAOJ). This changed, however, and so, in 2005, we were able to take the next step by creating the East Asian Core Observatories Association (EACOA) with the national astronomical Institutes of China, Japan, Korea and Taiwan as members. Building on this, in the 2006 EAMA symposium, I proposed a next step for the future: The establishment of an East Asian Observatory (EAO). While I always saw ESO as the most desired organisational model, I thought that it would not be possible to duplicate it in East Asia. After all, the political situation in Asia is different from the one in Europe. But we now have an organisational setup based on a collaboration between national institutes and—thanks to Paul Ho of ASIAA, Taiwan, and Masahiko Hayashi, the current NAOJ Director General, and others—the EAO was successfully established by ASIAA, KASI, NAOC, and NAOJ as partners. Though the size is still small, the EAO has received a concrete focus with the operation of the James Clerk Maxwell Telescope in Hawai'i. Thus cooperation in astronomy in East Asia has intensified over the recent years, supported by the fact that the level of astronomy in several Asian countries has clearly risen significantly.'

This is also bound to strengthen the IAU. Indeed spreading its activities globally, with the Regional Meetings and the OAO and OAD offices become a self-enhancing process.

'The main part of what appears to be a constant increase in the IAU national members since the 1950s has come from "moderately developed" regions and regions "under development". Especially after 1990, the increase of IAU individual members in the Asian-Pacific region is prominent, reflecting the growth of the economy and scientific activities in this region. IAU has a long history but its national members still remain around 70. We have many more countries that study and contribute the modern astronomy in Asian-Pacific, Latin America, and Mid-East and African regions. Promotion of astronomy in these regions, accompanied with the OAD and OAO, is therefore the way IAU should go. At the same time, astronomy is becoming more and more regional and also global. In Japan, I am encouraging our young astronomers to become involved in the IAU, e.g. by proposing symposia.'

Kaifu retired from his post as Director General of the NAOJ in 2006, but has remained very active in astronomy and science in general, having later served on

(Continued)

numerous important committees aside from his IAU involvement, e.g. the Science Council of Japan, as Chairman of Japan's government committee 'Large-Size Project of Sciences', and review committees for a number of astronomical institutes etc., providing advice and maintaining his vision about international cooperation in astronomy. In this, he is no longer a lone figure in Japan.

The Strategic Plan 2010–2020

The lessons of the IYA were too important to leave it at this. Many of the activities cried for extension or repetition, newly developed resources (especially online) needed to be preserved, and it was especially valuable to retain the global network of outreach and amateur organisations. This was the goal of the IYA 'legacy programme'.

IAU Vice President George Miley wanted to be even more ambitious, however. He especially focused on the development aspect of astronomy. He was the initiator of the Universe Awareness programme for economically disadvantaged children of primary school age, which had been officially launched in 2006, and which had been nominated as one of the cornerstone projects of the IYA. He realised that funding for well-planned and sustained educational initiatives *could be* found, if one had an ironclad case—which the IAU had, based on its long and distinguished record of successful educational and development activities. This was the central idea behind the IAU Strategic Plan 2010–2020 '*Astronomy for the Developing World*'.

Already before the start of the IYA, in January 2008, the IAU started to reassemble its scattered, yet multi-level educational activities into one comprehensive set of initiatives. After iteration with the relevant international organisations, such as the United Nations Office for Outer Space Affairs, the Committee on Space Research (COSPAR) under the International Science Council (ICSU), and the International Union of Radio Science (URSI), the Strategic Plan was approved by the IAU General Assembly in Brazil in August 2009. In August 2012, the IAU General Assembly in Beijing approved an updated version of an even more ambitious strategic plan and its implementation for 2010–2020, more simply titled '*Astronomy for Development*'.[12]

---

[12] It is available from the IAU web site: https://www.iau.org/static/education/strategicplan_2010-2020.pdf

Voting at the IAU GA in Rio de Janeiro, 2009. Photo: IAU Archives

The aim of the 2010–2020 Strategic Plan was to use astronomy as an inspiration to advance science, technology and education in developing countries. It argued that astronomy provides an appealing entrance to science, and is a relatively cost-effective way to start advanced scientific research, because many instruments can be used by guest observers, and much data is freely available. Besides, astronomy is a driver of technical innovation in optics, electronics, and other fields. So the core idea was not just to develop astronomy, but to use astronomy for development in a much broader sense—hence the title '*Astronomy for the Developing World*'.

The Strategic Plan called for sustained, long-term activities, building on the experience of the IYA as well as on existing activities, such as Teaching Astronomy for Development (TAD) and the programme group for world-wide development, to stimulate development of primary, secondary and tertiary education, as well as science research and the public understanding of science. This was an effort far beyond anything the IAU had done before. Since the IAU's own staff has always been small, expansion would be needed as well—including the extra funding that this required. For this reason, a key step in implementing the plan was to create a central coordinating IAU office of '*Astronomy for Development*' (OAD). Fortunately, this proved possible with extra funding from the member countries as well as external funds.

After a lengthy international selection process, the IAU chose South Africa and the South African Astronomical Observatory (SAAO) to host this office, with the South African National Research Foundation as the host institution and strongly supported by the Department of Science and Technology. In April 2011, the new OAD in Cape Town was inaugurated by the South African Minister of Science and Technology, Grace Naledi Pandor, with Kevin Govender, one of the primary organisers of the IYA, as its director. It supported activities around the world, stimulating astronomy teaching and outreach as well as providing institutional support.[13]

Since its foundation, the OAD and its small staff have already mounted an impressive programme on a very broad front. They have been particularly successful at mobilising astronomical communities around the world. A world map is shown on the OAD web site, highlighting the wide geographical diversity of its activities. The OAD centred on its three Task Forces (Universities & Research; Schools & Children; and Public) and ten Regional Offices & Language Centres in all parts of the world. The newest, the European Regional Office, was inaugurated in The Netherlands as recently as 26 February 2018, and will be operated jointly by the European Astronomical Society and Leiden University. This illustrates that the scope of its activities now reach further than the developing world.

---

[13] The OAD web site (http://www.astro4dev.org/) gives a constantly updated overview of its activities.

The IAU Strategic Plan 2010–2020

In 2018, the General Assembly approved a new, even more comprehensive Strategic Plan 2020–2030 that builds on these ambitions, connecting them to the UN Sustainable Development Goals. One of its new plans is a new Office of Astronomy for Education, to stimulate and support astronomy teaching at schools of all levels.

Other outreach activities were also expanded and professionalised after the International Year of Astronomy. The Office of Astronomy for Development was the first professionally run IAU office apart from the Secretariat in Paris, but soon after, several others were established, always with external support. In 2012, the Office for Astronomy Outreach was established in Mitaka, Tokyo, Japan, in cooperation with the National Astronomical Observatory of Japan (NAOJ), to coordinate the follow-up of the International Year of Astronomy, especially to maintain and expand the network of outreach organisations.

In 2015, the IAU signed an agreement with the Norwegian Academy of Sciences and Letters, creating the IAU Office for Young Astronomers (OYA). Chaired by an IAU Vice-President (currently Liu Xiaowei), the main task of the OYA is to serve as a robust financial and organisational basis for the operation of the International Schools for Young Astronomers. This secured the future of this half-century-old IAU 'Flagship Programme', which by now has trained c. 1500 young astronomers over the years, at least one third of them women. In fact, in some years, more than half of the participants were women. The Norwegian Academy now sponsors one School per year, which makes it possible to organise a total of three Schools every two years (there are plans to increase this frequency). Currently, Kam Ching Leung (USA) serves as ISYA Director and Itziar Aretxaga (Mexico) as Deputy Director.

One of the IYA cornerstones, Astronomy & World Heritage, led to an agreement with UNESCO in 2008 to establish the Astronomy and World Heritage Initiative, for which the IAU set up a dedicated working group. It did not only endorse the dark sky as the heritage of all humanity, but also more specific sites where that sky had been studied, from Ulugh Beg's observatory in Uzbekistan to the radio telescope at Jodrell Bank.

Finally, one way to make the IAU, and astrophysics in general, more visible, was to cooperate in awarding prestigious prizes. In 2000, the Gruber Prize in Cosmology was awarded for the first time. The IAU agreed to nominate some of the jury members, but the Gruber Foundation also founded two fellowships for young astronomers at the same time (see interview with Andersen). A few years later, the Kavli Foundation founded three major prizes for fields that were not covered by Nobel Prizes, including one for astrophysics, although several Nobel Prizes have been awarded to astrophysicists. Again, the IAU cooperated by advising the Norwegian Academy of Sciences on the nomination of the prize committee members. The Norwegian Academy and the Kavli foundation also support early-career researchers, in this case via the International Schools for Young Astronomers. And from 2016, each IAU Division also offers a prize for the best PhD thesis in its field.

The centenary in 2019 will be used as an occasion for a new round of outreach activities, building on the experience of the International Year of Astronomy. The 2018 General Assembly in Vienna already featured projects for people with special needs, including visually impaired people, an open-access exhibition, and activities to commemorate the 1919 eclipse observations that provided the first proof of the General Theory of Relativity. The founding of the IAU itself will be celebrated in 2019, in the Palais des Académies in Brussels, where the Union was founded 100 years ago.

Membership and Administration

Despite the increasing scale and scope of the IAU, its central administration remains very small. In practice, the (unpaid) General Secretary is assisted by a staff of only three or four people for administration, finance, database management, editorial assistance and archiving. The Executive and Financial Committees can supervise the secretariat, but from a relative distance.

Otto Heckmann and the Nobel laureate Gerhard Herzberg at the IAU General Assembly in Hamburg, 1964. Photo: ESO Archive/W. Seggewiß

The office receives help from other astronomical organisations, especially ESO, which has its headquarters in Garching in Germany. The ties between the two organisations had always been close, with for example several (former) ESO-officials serving as General Secretary or President, including Blaauw, Woltjer, West, Andersen and Corbett. Otto Heckmann and Catherine Cesarsky were even President of the IAU and ESO-Director General at the same time! In the first decade of the twenty-first century, the connections were again especially close. ESO staff members effectively handled much of the public relations of the Union. A central figure was Lars Christensen, who acted as IAU press officer from 2006 (starting during the tumultuous Pluto debates). Thanks to Cesarsky, the IYA secretariat was also hosted by ESO.

The General Assemblies kept growing; c. 3400 participants attended the General Assembly in Vienna in 2018. But while previously the Executive Committee sometimes worried that organising them was too large a task for local committees, now each General Assembly is subject of a highly competitive bidding race between several cities. Hosting a General Assembly is now a prestigious honour!

The General Secretary and one Vice-President each oversee the dedicated offices for Development, Outreach, and Young Astronomers. Of the specialised 'service' bureaus, only the Minor Planet Center remains, mostly funded by NASA. The Central Bureau for Astronomical Telegrams moved to the Department of Earth and Planetary Sciences at Harvard University in 2010; since 2015 it is no longer supervised by IAU Commission 6 (which was abolished). The time and latitude bureaus had already moved out of the IAU's control in the 1980s.

But while the IAU became a vibrant, international organisation with a significant budget, the way of working at the office was still that of a small, relatively informal organisation. This went well for a long time, thanks to the dedication of the staff and the high level of personal trust, but in long run it was untenable. This is the context in which the 'Orine affair', which several of the interviewees in this book referred to, should be understood.

For twenty years after 1987, Monique Léger-Orine was the executive assistant at the office in Paris, in charge of the day-to-day running of the affairs of the Union. All General Secretaries who worked with her praised her efficiency. Characteristically for a small-scale organisation, she was the one and only person who knew how everything worked. But soon after her sudden, unexpected death in January 2008, only months before her retirement, her successor Vivien Reuter discovered a number of major irregularities in the financial accounts. It finally emerged that Orine had embezzled several hundred thousand euros of IAU funds. Since there was no criminal investigation (she was no longer there), the background was never explained. Personal gain appears not to have been her motive; according to some people, she may have tried to help people close to her, who were in severe personal trouble. The affair was duly reported to the General Assembly in 2009.

For the IAU itself, this incident revealed how much the working procedures had been based on personal trust and informal consent. Successive General Secretaries have reorganised the administration and introduced more modern management tools, but the office still remains small for such a wide-ranging organisation. On the positive side, the longer-term financial situation of the Union was actually better than expected, since a significant share of the operating costs had been fictitious. Moreover, part of the damage was recovered from the bank, which had not followed proper procedures (a lawsuit against the auditor was lost).[14] The recovered funds were used to support the Strategic Plan 2010–2020.

[14] IAU Archives box 1B, file 'Officer's meeting 15–16 Jan 2015'.

Thierry Montmerle: From High-Energy Physics to Astronomy, from Organisational Reform to Indigenous Protests

Assistant General Secretary (2009-2012), General Secretary (2012-2015), Chair of WG Public Naming of Planets and Planetary Satellites

Interview conducted in Paris on 29 June 2017

The IAU has benefitted from members and officers who showed great dedication to and deep involvement in the Union. Thierry Montmerle was no exception, but his engagement was originally spurred by frustration. He transformed this into an important contribution to modernise the Union, serving at a time when huge changes in astronomy were in the air.

Thierry Montmerle at his home institute, the Institut d'Astrophysique de Paris. Photo: C. Madsen

'I was interested in astronomy since I was a child, however, my family didn't include any scientist. I read astronomy books. It was a classical story and yet, I did not choose the straight path into astronomy as a job. Space fascinated me, too. I was 11 when the news about Sputnik broke. It was as if a lightning had struck me. Later I embarked on scientific studies at the École Normale Supérieure in Paris. I was interested in physics and it included an option in astronomy, which I did at the Institut d'Astrophysique in Paris with Evry Schatzman as my mentor. I did my "*Doctorat de 3e cycle*" degree with Schatzman, and also with his close colleague Jean-Claude Pecker.

Schatzman who drew my attention to the matter/antimatter research in the early 1970s. The first scientific results from space regarding the gamma-ray background had come in and it was suggested that an intriguing "bump" in the gamma-ray spectrum could be caused by a red-shifted annihilation process, taking place a few million years after the Big Bang. Also, quasars had been discovered a few years earlier and some people thought that as their energy source annihilation processes might be at work. Unfortunately, I soon realised that it was a dead end, but my curiosity was aroused, so I continued to study the gamma-ray background by introducing the idea of "cosmological cosmic rays", which could produce gamma rays through collisions with matter as a result of

(Continued)

supernova explosions in first-generation stars. Then, at Saclay, I came into contact with Hubert Reeves, who at the time was exploring the origin of light elements and with colleagues had demonstrated that these elements were the product of collisions of cosmic rays and interstellar matter. Hubert Reeves became my PhD supervisor. Coming from Quebec, he also helped me to do my national service at the University of Montreal, where I stayed for two years working with Georges Michaud, who studied element diffusion in stellar atmospheres. This brought me "back to the realities of stars", even if my first interest remained high-energy physics and satellites. Returning to Saclay, a group there worked on COS-B, launched in 1975, which was the ESA satellite that drew the first complete map of the gamma-ray sky—it took ten years to complete. This is where I made the link between the physics of cosmology and stellar physics because I discovered that some gamma-ray sources in the galactic plane were correlated with massive star-forming regions hosting supernova remnants. I later moved on to star-formation and early evolution processes as revealed by their X-ray emission, and ultimately, while I became Director of the Laboratoire d'Astrophysique de Grenoble in the early 2000s, to low-mass stars, their accretion discs and planet formation. In the end, this scientific adventure really was for me a series of lucky episodes—opportunities with great people—starting with early contacts with physicists at CERN. It was like being dragged into an entirely new and fascinating era. Everything was new, space astronomy was at its beginning. For me, the driving force was a mixture of emotion regarding space technology and astronomy of the "invisible Universe" itself.

To be honest, it is out of frustration that I became involved in IAU matters. As a young researcher, I wanted to organise an IAU symposium about gamma-ray sources and star formation, but the application was rejected. My frustration was not caused by the rejection, but because I saw a lack of links between some of these "new" research areas and the IAU. Later I became involved in the IAU High-Energy Astrophysics bodies, but not in a deep way.

At Saclay, most of us were physicists, not astronomers. In some ways, that was a drawback because we felt a bit cut off from mainstream astronomy. But the gamma-ray sources were the first objects in this high-energy range for which you could expect to find a counterpart at other wavelengths (after X-rays). So we were coming from physics to astronomy, yet I felt that we were not in the right circles to have a successful impact at the IAU.

Also, the Division structure itself seemed in need of change. For example, in 1997 Division XI "Space and High-Energy Astrophysics" was created—and I was a member of its Organising Committee. Yet it was identical to the previous Commission by the same name, which continued to exist, and it had no other Commissions. In addition, there were oddities like two Divisions dealing with stars—"Stars" and "Variable Stars". A "mysterious" Division XII called "Union-Wide Activities", added in 2003, was essentially a basket where those topics had been put that didn't fit into the established pattern, even though they were as different as Education and Databases; etc.

In my view, and perhaps more importantly, the IAU was a great organisation, but it somehow missed the point of the evolution of modern astrophysics. This appreciation was in fact shared by others, who felt that it was a waste of intellect that the IAU was not thoroughly represented in some of these new areas of astronomy. For me, this frustration actually continued until I was encouraged to

(Continued)

become involved in the management of the Union. But frustration can be a very good driver, provided that you can something about it. Yet, despite this state of mind, for a long time, I did not have any idea to transform the IAU.'

As of 2009, Montmerle took note of the new thrust of the IAU regarding public outreach, not the least connected with the IYA and its follow-up activities.

'It was, of course, a necessary endeavour, but I had the feeling that along the internal lines, the IAU itself, too, needed reform. Inside the IAU everything looked to me somehow frozen. To simplify, there was the "real" astronomy—the work done in the institutes and the agencies, like ESA and ESO in Europe—there were the tremendously successful IAU outreach activities, and then, in between, there was the IAU that I felt was disconnected from the dynamic world of modern astronomy and astrophysics.'

To Montmerle, the need for reorganising the IAU Division structure had become evident. When was approached by Catherine Cesarsky, then President-Elect, about the possible appointment as IAU Assistant General Secretary, it gave him a chance to tackle this challenge.

'Sometimes, ignorance can be useful. Had I known in detail how the IAU functions, I probably wouldn't have broached the topic of reorganisation of the Division structure. The role of the General Secretary is primarily to look after the finances, to keep the Symposia running, to organise General Assemblies, to manage the house and so forth. But I didn't fully realise it, and so at the first Executive Meeting organised by Bob Williams as IAU President and with me as Assistant General Secretary, I showed a few slides suggesting a new divisional structure, and the Executive Committee invited me to elaborate a formal proposal for reforming the Divisions. It involved a lot of technical, quasi-legal, aspects, for which Ian Corbett as General Secretary kindly provided his expertise. We also reviewed the documents from the time, when Lodewijk Woltjer introduced the original Division structure, which was approved by the XXIIIrd General Assembly in Kyoto, twenty years earlier. Furthermore, the restructuring work involved the Division Presidents and I had the privilege of working with a wonderful set of Division Presidents. Not that it was easy since unsurprisingly, changes often lead to resistance and it certainly did within some of the past Divisions.

The heart of the proposed new organisation, while apparently administrative, was really scientific in the sense of providing the IAU with a structure that would enable its members to do science through the IAU, not simply produce reports. So the goal was to create fora in which they would feel encouraged to meet and talk because they would share a common scientific goal. The new, more concise structure also made the activities of the IAU more broadly understandable to the public (and to the member countries funding the IAU)—in particular by the creation of a new Division on Education, Outreach, and Heritage.'

The number of Divisions was reduced from 12 to 9 and new designations were introduced, the Divisions now being known by a letter, not anymore a roman number. In the end, the new structure was adopted by the General Assembly in Beijing in 2012.

'A further challenge was to organise the election of the new Division and (later) Commission officers. Historically, the IAU has not displayed much of a lively and dynamic membership democracy, though formally, elections did take place. In reality, too often it meant "sorting out things" among a small group of people—engaged and dedicated, no doubt, but somehow also perpetuating

(Continued)

themselves. This made at least some people quite frustrated, and we wanted to change this. So the challenge was to have members stand as candidates, asking them to develop and present their election platform and then to have members take an active part in the vote.

After the restructuring of the Divisions came the changes of the Commissions—no less daunting, in fact. This was implemented during my term as General Secretary, but of course in close collaboration with the newly elected Division Presidents. It was closer to the broad membership of the Union and therefore also viewed more critically. We wanted people to blend into the new structure, select their Commissions and working groups and create new working groups where necessary. The idea was very much to enable the accommodation of new ideas, so we created flexible structures with a finite lifetime that, however, could be re-established, should the need arise and simply approved by the next hierarchical level.

The final blessing happened at the Honolulu General Assembly in 2015. Here, among other novelties it was gratifying to see the establishment of a Commission on Gravitational Wave astronomy, a few months before these were detected, and I believe that it is the first time that the IAU has been ahead of the science, not simply following it. Also, a new Commission for Computational Astrophysics is now in place for this emerging field—large computers could be understood as an alternative kind of telescopes to investigate the universe.

In any event, I'm truly indebted to the Division Presidents, but I also wish to acknowledge IAU Presidents Bob Williams and Norio Kaifu, whose encouragements and "go-and-do-it" approach was decisive and something that I immensely appreciate.'

It was a time with many other activities, not the least the IAU restructuring and the implementation of the Strategic Plan, including the establishment of the Office of Astronomy for Development (OAD) in Cape Town and the Office of Astronomy Outreach (OAO) in Tokyo, the latter at the initiative of Norio Kaifu.

'The naming of exoplanets initiative, implemented by the OAO, was strongly supported by Bob Williams and IAU Press Officer Lars Lindberg Christensen. In a sense this was a fall-out of the Pluto debacle, especially taking into account the strong and very negative reactions in the USA. These reactions were not confined to the broad public but included professionals involved in NASA's "New Horizon" space mission to Pluto. In fact, the principal investigator, Alan Stern, on behalf of the mission, refused to "recognise" the IAU decision to re-classify Pluto as a Dwarf Planet, and later launched his own initiative, the for-profit undertaking Uwingu, calling in the public to freely name Mars craters and also exoplanets, thus openly challenging the authority of the IAU to name celestial objects.'

The issue was critical because the IAU has historically assumed the role of the world body that assigned official names and designations to celestial bodies. Critics have argued that the sky "belongs to everyone" and that the naming role of the IAU is self-appointed.

'However, the naming of celestial objects was part of the original tasks of the IAU. For me, the formal authority of the IAU in this matter is derived from the fact that within the tri-annual reports to the General Assembly, the names are approved by the official delegates of the national members of the IAU, who represent the individual academies of science or similar scientific bodies.

(Continued)

More profoundly, naming is really a scientific activity. The naming of Moon Craters is a prime example, where before the IAU was created, confusion existed because people looked at the Moon with different telescopes with different resolving powers and found it difficult to agree when they discussed their observations of individual objects having different names. To deal with this key issue, a Commission dedicated to the Moon was in fact created as part of the IAU in 1919. Also during the "Moon Race" of the sixties, we saw instances where American and Russian space agencies wanted to give different names to the same features on the Moon—the Russians being the first to observe the far side of the Moon. Here the IAU played an important role in having the American and Russian astronomical communities negotiate and eventually agree on their designations.

Having a body like the IAU dealing with this issue also means that there is a commonly accepted naming process with clear criteria and rules. We have indeed seen with dismay private initiatives to sell planet and star names, which we find unacceptable. In fact, the IAU has even been consulted by the French Ministry of Finances about complaints from the public concerning selling names of celestial objects in France. Be that as it may: The IAU naming of extra-solar planets initiative—the so-called "NameExoWorlds" contest was our attempt to tackle this difficult issue by maintaining the scientific rigour and a transparent, free process while also reaching out to the public.

The contest involved collecting name proposals from the public, whereby we involved the large body of amateur astronomers across the world. The public voting process on the proposed names was launched at the General Assembly in Honolulu in August 2015. The outcome was great in the sense that we had half a million votes for naming 19 exo-planetary systems selected by the IAU (with up to 5 planets), as well as their host stars, but we also saw some pitfalls and the amount of work involved in such an initiative and so, it's an open question how this will evolve in the future. Still, I really think that the centenary of the IAU in 2019 would constitute a great opportunity to revive this event.'

(Continued)

Lisa Kaltenegger and Thierry Montmerle launching the public exoplanet naming initiative at the IAU GA in Honolulu. Photo: C. Madsen

To complement the Office of Astronomy for Development (OAD) in Cape Town and the Office for Astronomy Outreach (OAO) in Tokyo, in 2015 the IAU established an additional Office for Young Astronomers (OYA) in Oslo, capitalising on the long-standing International Schools for Young Astronomers (ISYA) supported by the Norwegian Academy of Sciences and Letters, but with plans for expanded activities in the future.

'The very successful ISYA activities have, in my view, not been sufficiently publicised. Thanks to Oddbjørn Engvold, this office has now been established in Oslo. It also underlines the fact that the IAU is not "just" in Paris, but that it is active across the whole world.

Another initiative that I worked on as IAU General Secretary was the suggestion to include sites of astronomical significance to the UNESCO list of World Heritage Sites. IAU already had a cooperation agreement with UNESCO, signed by Karel van der Hucht, that we could build on. The first world heritage site that was approved was Wilhelm Struve's Pulkovo Observatory in St. Petersburg. This is obviously of historical importance, but for UNESCO it is not necessary to be a historically important place, but it has to be important for knowledge and important even for the present-day public. So an active observatory—or observatories—can be put on the list because of its importance for humanity. From the IAU, it is primarily Clive Ruggles, who as our expert has drafted a number of "case studies" in support of this collaboration, and created the "UNESCO-IAU Portal to the Heritage of Astronomy".'

Without prejudice to the many exciting scientific presentations, to the uninitiated, IAU General Assemblies will often appear as fairly staid affairs. But not always. The 2006 meeting in Prague that struggled with the hugely controversial topic of Pluto created a worldwide public echo. Nine years later, the General

(Continued)

Assembly in Honolulu saw itself mired in another controversy, in a way a local Hawaiian conflict and yet one that would leave no astronomer, regardless of nationality, untouched or indifferent. And yet, the IAU chose to stay above the fray.

'It was a difficult General Assembly. From a science point of view, I believe that Honolulu unquestionably was a success. The programme was good with many good symposia. The problems we faced, however, were manifold. One was cost, partly because of changes in the exchange rate with the dollar having risen considerably since the decision to have the General Assembly in Honolulu was taken. But one reason to select Honolulu, six years earlier, was obviously the presence of the observatories, especially the one on Maunakea, one of the world's premier sites for ground-based astronomical observations. And then, during the General Assembly the public excursions to the observatories had to be cancelled due to roadblocks on the way up the mountain.'

For years, the Maunakea observatories have been at the centre of an increasingly bitter conflict between groups of native Hawaiian activists, for whom that volcano is a sacred site, and the observatories operators, mainly US universities but also with major international partners. Before the IAU General Assembly, the planned Thirty-Metre Telescope (TMT) by the California Institute of Technology with partners from Canada, China, India and Japan, had begun the construction on Maunakea, only to be met with a relatively small, but determined group of protesters. Aside from the issue of the sacred mountain, in fact, the various groups of activists field a multitude of grievances towards the USA, from historical acts of suppression to complaints about environmental destruction today. The TMT, in essence, became the symbol and lightning rod of long pent-up anger among many an islander.

Protesters outside the conference centre hosting the IAU GA in Honolulu, 2015. Photo: C. Madsen

(Continued)

'We feared to have huge anti-TMT demonstrations in front of the conference hall. There were threats to that effect. We, the IAU and the US organiser, the AAS, had meetings with the authorities of the City of Honolulu and the State of Hawai'i as well as hiring a company to act as a go-between with the various groups of activists in order to avoid a major confrontation. Eventually, there was no major demonstration, aside from a peaceful march through Honolulu on the Sunday in the middle of the General Assembly with perhaps 10,000 participants, but protesting again a variety of local issues. At the conference itself, a couple of pickets held out, engaging with many a conference participant, albeit in a civil, almost friendly, atmosphere. It helped enormously that Norio Kaifu, the IAU President, knew well the local community from his time as Director of the Subaru Telescope Project and he knew some of the activists, too, so he could play a very positive role.'

Upon arrival, the conference participants received an introductory note, prepared by the IAU and the AAS, explaining the situation and also expressing that this being a local conflict, the IAU did not wish to take a side. For the same reason, the wish of the activists to address the General Assembly was denied. And yet, a representative of the activists was given the floor during a public talk on one evening.

'It was difficult, because for the IAU not to defend one of the most important telescope projects of the future might seem embarrassing. It was not the happiest experience, but at least things stayed peaceful and, seen from the participants, the General Assembly could take place with no major problems.'

Piero Benvenuti, Seasoned Manager, Dedicated Institutional Reformer

Assistant General Secretary (2012–2015), General Secretary (2015–2018)
Interview conducted in Paris on 17 May 2018

As the IAU approaches its 100th Anniversary, fresh winds are blowing through the organisation. With a surge of new member countries, an ambitious new Strategic Plan and new initiatives to rejuvenate the membership body, the IAU has been put on an exciting track into the future. One of the engineers behind this was Piero Benvenuti, the IAU General Secretary 2015–2018.

(Continued)

Piero Benvenuti at the IAU Office in Paris. Photo: C. Madsen

'I got interested in astronomy very early on in my life. I remember getting an old theodolite with an objective of 4 cm, like the Galileo telescope, building the mount in wood and looking at the stars with it. In high-school, I was attracted by Physics and Mathematics and listening to a public talk about the state of affairs of astronomy in the late 1950s/early 1960s, when the theoretical studies of the structure of stars were becoming mature, giving insights to the evolution of stars, their colours etc.—something that I could see with my little telescope—set me on the path towards astronomy.

I enrolled at the University of Padova and when I did my thesis, I spent several months at the Asiago Observatory that at the time hosted the largest optical telescope in Italy. For my thesis, I used the Asiago Schmidt telescope with multiple filters to detect objects with an excess in UV-light and produce a catalogue of very blue starlike objects in the Coma field.

I got my degree in 1970 and following military service, I joined the staff at Asiago.

It was an exciting time with huge scientific progress built on technological advances, as this book also describes: computing facilities were becoming available, and the image intensifiers and soon thereafter solid-state detectors came about. Space astronomy was emerging as a new window on the Universe. For example, I remember listening to Riccardo Giacconi, presenting the first results from the UHURU mission. It was mind-blowing with X-ray sources everywhere in the sky, so I got interested in space instrumentation. That was also at the time when the International Ultraviolet Explorer (IUE) was launched. In many ways, the IUE was breaking new grounds—it really attracted the astronomical community to space astronomy, because it was a real observatory. Also, it introduced an operational mode that would later become standard also for ground-based observatories. The IUE furthermore provided a digital archive that could be used not just by the observers themselves but also by other astronomers. ESA was looking for resident astronomers to work at with the IUE and I got the position, later becoming the IUE Observatory Controller.'

(Continued)

Benvenuti stayed in Madrid for eight years and then became Head of the joint ESA/ESO Space Telescope European Coordination Facility. In 2003, he resigned from ESA to become Commissioner, then President for INAF, at the time a newly created institutional structure for Italian astronomy, with further senior managerial assignments to the Italian Space Agency (ASI). In 2012, he was elected as IAU Assistant General Secretary, taking over as General Secretary in 2015.

'I joined the IAU in the early 1970s. My professor at Padova, Leonida Rosino, was quite keen for his people to become IAU-members. Even so, my "institutional involvement" is of a more recent date. In fact, it was Bob Williams, whom I, of course, knew from the Space Telescope Science Institute, who asked me if I would be willing to serve as IAU Assistant Secretary General. I was still teaching at the university at the time, but I would be free three years later for the term as General Secretary, which is really a full-time task. And so, at the General Assembly in Beijing, I was elected. For the first two years, my main tasks comprised the selection of symposia and taking care of publications, but gradually, I become more involved in the overall tasks of the office, including some aspects of the Strategic Plan implementation relating to the OAD in Cape Town. More importantly, I became interested in the financial affairs and the usability of the IAU database. In any event, this was useful when I took over as General Secretary.'

The 'Orine affair' had taught the IAU not to take its eyes off the ball when it came to financial matters. That included the institutional members and thus at the General Assembly in Honolulu, the IAU Executive realised that renewed strengthening of the operational practices had become necessary.

Entrance to the IAP and the current IAU Secretariat in Paris. Photo: C. Madsen

'The IAU office is really a small operation, and the way the IAU functions, mutual trust is a key requirement. But of course, the trust also extends to the IAU-Member Countries' relations. With this in mind, when I took up my duty as General Secretary, I worked full speed to improve the transparency and reliability of our financial management, partly by implementing new accounting system

(Continued)

software, which in turn was enabled by the hiring of new office staff, and by expanding the usability of the database. The database was tailor-made and it is of course run by ESO. But I wanted us to be able to use it more freely, without constantly involving ESO, and also link it to the accounting system. So, while history taught us that vigilance is prudent, a continuously evolving IT system has enabled us not only to ensure full transparency regarding our financial transactions, but also to alleviate the administrative burden on the office staff somewhat, which in turn has allowed them to spend time in support of more creative tasks. On top of that, we've actually able been able to save money. The combination of efficient IT systems and truly dedicated staff is key to success, and we have both!'

At the General Assembly in Honolulu, it was decided to prepare a new Strategic Plan for the Union for the 2020–2030 period and Benvenuti joined a small working group with Debra Elmegreen, Ewine van Dischoeck and Renée Kraan-Korteweg.

'Of course, for me as General Secretary, much effort has gone into the implementation of the Strategic Plan from 2009, but indeed also to think about the follow up with a broader and much more ambitious plan for the coming years. I feel very passionate about the activities and the perspectives in the Strategic Plan, and one of the visible results of the comprehensive worldwide outreach and educational initiatives—from the OAD system of regional nodes to the ISYAs—is that we're now expecting a surge of new member countries, though a few are re-joining. The list includes Nigeria, Morocco, Algeria, Jordan, the United Arab Emirates, Mozambique, Madagascar, Slovenia and Syria.

It's interesting that most of these countries are Muslim countries and I believe that the Islamic role in the history of astronomy has been much more substantial than what we see today in history books, and this should be recovered.

My only regret is that I did not manage to get Palestine as a national member. Suleiman Baraka has done a lot of work in Palestine—working with the OAD, especially for children in the Gaza Strip, but so far, it has not been possible to move to real IAU membership.

Piero Benvenuti (left) and Ethiopian Minister of Science and Technology Demitu Hambisa (right) signing an agreement to host an East African regional node of the IAU Office of Astronomy for Development. Photo: IAU/OAD Archives

(Continued)

The 2009 Strategic Plan was a huge leap forward for the IAU that in many ways gave new impetus to the Union, perhaps even changing its thrust. The new Strategic Plan builds on the experience of this and, one can say, cements the legacy of the International Year of Astronomy. The inclusion of education is one example. And yet I believe that the new plan marks a step no less important than what we took in 2009, offering a complete description of the IAU in its entirety and with clear indications of the future development.

Overall, as regards the education aspects, what we're trying to achieve is a more structured way of dealing with astronomy teaching at university level, where we're providing high-level courses in countries where these courses do not exist already and we wish to increase the number of these courses. This will be organised by the Office for Young Astronomers, mainly through the joint IAU/NAS ISYAs. As regards education, i.e. below the university level, we're now proposing to create an additional office for education so that we can incorporate the wide range of existing activities, but within a more coherent system. So now we will have the OAD for Development, the OAO for Outreach, the OYA/ISYA for astronomy teaching at university level and the OAE for education.

I hope that the new Strategic Plan will serve as a guide for future IAU officers, just as the 2009 Plan has been in the preceding period.'

The IAU will of course not neglect its other functions and services to astronomy for which the Union was created, but as times change, so does the context in which these activities take place.

'As regards future astronomical facilities, the IAU constitutes a forum where people can express themselves freely and this enables the IAU to develop a neutral, project-independent vision on astronomy. Another aspect of "promoting astronomy in all its aspects" is to help those countries without a strong astronomical community to develop. One way that we are pursuing is the inclusion of the junior members—a new member category of which we have great hopes. We have created a working group for junior members, to help us understand better their professional and societal situation and to propose actions that can support them. This may also help us with another burning issue: The gender problem. Firstly, alongside with the Working Group on Women in Astronomy, the group may help us also understand the specific challenges facing young female researchers. Indeed, the barriers, some of them seemingly invisible to us, have to be brought into the open so that we can tackle them. Secondly, we see a considerable rise in female memberships in this particular category. So the opening of the Union to junior members may also help combat the gender imbalance.'

At the General Assembly in August 2018, Benvenuti's term came to an end, though he will continue to help specific projects that he has encountered while in office, especially in the emerging countries.

'The job as IAU General Secretary has been really exciting. I especially enjoyed the possibility of interfacing with different cultures that I hadn't met before—where astronomy is the common denominator albeit conjugated in different ways. For example, I was surprised and impressed by the activity at semi-professional level in Iran and also a recent meeting in Jordan was very inspirational for me. It has also given me the chance to kindle my personal interest the relations between astronomy and other intellectual activities, e.g. religion in various parts of the world including the Islamic countries. I am grateful to have had the chance to learn more about this and it's something that I wish to cultivate after my time as IAU Secretary General.'

The IAU Reformed

Despite all these new activities, the more traditional activities of the IAU remain the core of the Union. In 2015, the IAU office reported that about 24% of the budget was devoted to education, 16% to science (supporting meetings, primarily Symposia, and publications), and 14% to the General Assemblies (including travel grants). The rest went to the office, communication (including the website, electronic Information Bulletin and, from 2006, a newsletter) and general administration.[15] This shows that organising and supporting meetings of the highest quality was still a major activity; most of the contributions of the National Members actually returned to the scientific communities, not to running the IAU itself.

Most of the meetings and publications, apart from the General Assemblies, are organised by Commissions and Divisions, which form the scientific backbone of the Union. The IAU name functions as a label of quality, stimulating scientific interest, but also boosting the prestige of attending or organising one, which could help obtaining political and financial support. This explains the continuing popularity of the symposium series. The colloquium series was terminated after colloquium no. 200 on direct imaging of exoplanets in Nice, France in 2005. After this, all scientific meetings would be symposia, which in turn received extra support.

As we have seen in the previous chapters, the number of Commissions and the division of topics had been subject of debates for many decades. To put it bluntly: for many, the Commissions provided engaging communities of like-minded people, but for others, they represented an ossified old-boys network of senior scholars, who carefully defended their own turf. In the twenty-first century, the commission structure was finally reformed. This happened in three steps: a revision of the Statutes and By-Laws to remove any formal obstacles to change, finalised in 2003; an overhaul of the Divisions in 2012, and finally the restructuring of the Commissions in 2015.

In 2003, a former, present and future General Secretary (Andersen, Rickman and Engvold) and a Vice-President (Williams) conducted a thorough revision of the Statutes and By-Laws of the Union.[16] The purpose was mostly to bring these documents up to date and weed out inconsistencies, but the revised texts also contained some reforms to encourage change. For one thing, the limit on the number of Commissions to which one could belong

---

[15] GA newspaper 2015, 30.
[16] They explained the new proposals in IB 94.

was lifted—one of the last remainders of the time when the goal had been to limit the growth of Commissions rather than stimulate it.

The 2003 rule that in the General Assembly, also scientific issues were voted by national members rather than individual members, was quickly reversed at the next GA, before the Pluto vote of 2006. The rationale had been that the national committees would prepare any vote by consulting their constituency, thus ensuring that a vote was based on the consensus of more people than the few hundred who attended the business meetings of the General Assembly. In practice, this rarely happened, however. Now the IAU mostly deals directly with the individual members, while the national members still vote on all issues with financial consequences (they pay the membership fees, after all!), and also play a crucial role in nominating individual members. A related long-standing legal anomaly was clarified in 2003: as a non-governmental organisation, the IAU cannot have 'member countries', only 'National Members' or 'adhering bodies', such as scientific academies or equivalent.

The biggest change involved the functioning of Divisions and Commissions. The Divisions were given more tasks, including supervising 'their' Commissions and 'monitoring their evaluation': coordinate the creation, merging, splitting or termination of Commissions. This was a serious job, because the Commissions were given finite lifetime. Until then, once founded, Commissions would automatically survive unless the commission members themselves decided to merge or terminate them. Twenty of the Commissions in 2000 had existed since the first General Assembly in 1922, with only minor adjustments in name (see Appendix B). The new statutes, however, included an automatic 'sunset clause': every three years, Commissions had to apply to be continued or be terminated by default (a first re-evaluation would happen after six years).

The new Statutes and By-Laws finally cleared the way for a thorough restructuring of the Commissions—in principle. In practice, it took another 10 years before it actually happened. Inspired by the success of the IYA, which proved that the astronomical community was actually capable of untraditional initiatives, incoming General Secretary Thierry Montmerle initiated a new attempt, strongly supported by the Executive Committee, which had returned to the topic of reform several times over the 10 years. This time it succeeded.

The commission overhaul was preceded by a reorganisation of the Divisions. When the Divisions were founded in 1994, they had been designed as umbrella structures for a group of existing Commissions (in 2003, a 12th Division was added for a last category of 'Union-Wide Activities'). The idea was that they would gradually encourage the Commissions to reform, but that had not happened. Even the 'sunset' clause formally introduced in 2003 did not change that: the Commissions had too much history, and the Divisions were too dependent on them. Now, the Executive Committee, after much debate, proposed a new structure with fewer Divisions, deliberately ignoring the existing Commissions in a top-down operation. Montmerle also searched for new division presidents, although they were later to be elected by the Division members.[17]

The new Division structure was confirmed at the General Assembly in Beijing in 2012. This finally cleared the way for the 'reset' of the commissions. Invoking the sunset clause, none of the Commissions was reconfirmed, meaning that they were all automatically terminated. In the three years up to 2015, new Commissions would be proposed. This led to a new structure (see Appendix B). Inevitably, some of the new Commissions resembled old ones, but several new topics were introduced. The most spectacular example was Commission D1 Gravitational Wave Astrophysics, which was founded several months before the first actual detection of a gravitational wave in September 2015 (and officially announced in February 2016). On the other hand, the number of Commissions on small solar system bodies was reduced, reflecting the change in the scientific focus of the astronomical community.[18]

The Divisions are now the prime internal sections of the IAU. This is reflected by the fact that all individual members must belong to at least one Division (there is no maximum). At the same time, a limit was again put on the number of Commissions of which one could be a member (3, or 4 if one of them is in Division C: Education, Outreach and Heritage). The idea that Commissions only should have active members has never completely disappeared.[19]

The General Assembly in Honolulu in 2015 reflected the more prominent role of Divisions: on top of the traditional Invited Discourses, Symposia, Joint Discussions (renamed 'Focus Meetings'), and Special Sessions, it featured one day of Division Meetings. In 2018, all Symposia also contained one plenary session to ensure that everybody could attend at least one meeting of each Symposium. This reflects the old ideal that General Assemblies should be occasions to look beyond one's own field of specialisation.

---

[17] Documents relating to this reorganisation are in IAU Archives box 14.

[18] Cf remarks by Montmerle in GA newspaper 2012, 71.

[19] These rules are specified in the Working Rules: https://www.iau.org/administration/statutes_rules/working_rules/

On the other hand, the social programme is now much less prominent than fifty years ago. There are still excursions and receptions, but no more concerts by top artists. The organisers of the General Assembly in Vienna did attempt to re-introduce dancing (waltzing, obviously!) at the closing ceremony—with mixed success.

The GA coffee breaks offer good opportunities for networking informal discussions. Photo: IAU Archives

Apart from the rearrangement of topics, another notable result of the Division and Commission restructuring was a significant increase in the number of female officials in the IAU: five of the nine new division presidents in 2015 were women. The Executive Committee also changed dramatically in this respect: in the history of the IAU, half of the women serving in the Executive Committee did so since 2008. Currently (2019), both the past and present President and the President-Elect are all women, as is the General Secretary.

Presidents' summit: IAU President Ewine van Dishoeck (2018–2021), former President Catherine Cesarsky (2006–2009), and former President Sylvia Torres-Peimbert (2015–2018) at the GA in Vienna. Photo: C. Madsen

The percentage of women is also steadily increasing among the common members; it now approaches 20%. Moreover, the women are very active: at General Assemblies, the percentage of female participants is routinely about 30%.[20] Dedicated meetings for women are also now standard elements of the conference; they are supported by the US National Science Foundation and the National Academy of Sciences. Also, childcare is now provided during the General Assemblies, and childcare costs can be included in travel expenses. In Sydney in 2003, childcare was provided because it was expected that some

---

[20] Cesarsky (2010); cf. GA newspaper 2015, 62.

astronomers would bring their family to Australia for the General Assembly.[21] Affordable childcare has been identified as one measure that would help women (and some men) to participate.

The active encouragement of young astronomers was underscored by the invitation of four Junior Members to address the closing IAU General Assembly in Vienna, 2018. Photo: C. Madsen

Since 2006, the General Assemblies include young astronomer lunches, where young researchers meet with senior scientists. They are sponsored by the Norwegian Academy of Sciences and Letters, which also sponsors the Office for Young Astronomers. There is also a 'consulting service' for early career researchers, and since 2018, recent PhDs can become 'junior members'. If they are still active in astronomy after six years, they can then move on to full membership—a recognition of modern career structures, in which the number of PhDs and postdocs is much larger than the number of senior research positions. 347 new junior members were accepted in 2018, one-third of whom are women. Membership is still by nomination only, and some nominations are still rejected.[22] Individual members pay no fee, so there are no financial impediments. And membership is for life.

---

[21] IB 93 (2003) 12.

[22] In 2015, 93% of the applications were accepted. GA newspaper 2015.

Silvia Torres-Peimbert, the Second Female Scientist at the Helm of the IAU and One with a Clear Mission

Vice-President (2000–2006), President-Elect (2012–2015), President (2015–2018)
  Interview conducted in Paris on 17 May 2018
  As the IAU approaches its 100th Anniversary, the Union is as lively and vibrant as ever with exciting plans for the future. It has fallen to Silvia Torres-Peimbert, as President of the IAU, to prepare the path leading into the Union's next century.

Silvia Torres-Peimbert in front of the Institute of Astronomy in Mexico. Photo: J. C. Yustis

Silvia Torres-Peimbert was born in Mexico City in 1940 and joined the National Autonomous University of Mexico for her studies.

'I started studying physics, not being particularly interested in astronomy; but in the early stages I took an astrophysics course and I simply loved it. There were very few people at the Observatory and very soon I was offered the opportunity of becoming an assistant at the observatory. On the staff was Guillermo Haro, the director, whom I admired greatly.

I did not have any specific academic interaction with Haro, but his personality prevailed at the Observatory. He shaped the whole structure of astronomy in Mexico. Another astronomer to mention is Paris Pişmiş, a Turkish scholar who arrived in Mexico in 1942 and remained there for the rest of her life. She, in fact, was the first astronomer with a PhD in the country.

(Continued)

Paris Pişmiş, here seen at the IAU GA in Grenoble. Photo: AIP Emilio Segrè Visual Archives, John Irwin Slide Collection

Haro was very keen for students to continue their graduate courses outside of Mexico. He lobbied very successfully to establish a scholarship programme for these cases. I was very lucky in receiving a scholarship to go to the USA. At the same time, I got married to Manuel Peimbert. Both of us were at the same stage of studies and we applied to Caltech and to the University of California in Berkeley. However, Caltech was an institution for men, and would only accept me as "the wife of a student", so Manuel and I decided to attend Berkeley starting in January 1963. It was a wonderful decision—not the least because of the diversity of studies offered there, including humanities and arts—with the open and broad attitudes that prevail in such an institution! Professionally, my path took me from star formation to Carbon stars, to stellar interiors, to inter-stellar matter—while geographically moving from Mexico to the USA and back again.

My interest in the IAU was stimulated by the fact that Haro had been an IAU Vice-President from 1961 to 1967; he was very proud of this achievement.

I attended my first IAU General Assembly in Brighton in 1970. It was awesome! I remember the Opening Ceremony, formally inaugurated by Margaret Thatcher, then Secretary of Education and Science in the UK. More importantly, as a new-comer I recall Martin Schwarzschild discouraging people from joining Commission 36 (Stellar Constitution) unless they had done significant work on that topic, so it was obviously quite prestigious to become a member of this Commission, and it was not open for any astronomer in the field. I found this attitude a bit disap-pointing, but that was the way of the IAU then. Indeed, it was perhaps to be expected for the "developed countries", but not so easy to understand for

(Continued)

astronomers from countries with less developed astronomy. Be that as it may, I became a member of the IAU and later I also joined Commission 36.

With the exception of the General Assemblies in Sydney and in Warsaw in 1973, I've attended every General Assembly of the IAU and I have fond memories of each of them. Of course, I remember the fascination of visiting New Delhi, the warmth of our colleagues in Buenos Aires and the beauty of Kyoto. In organisational terms, the General Assemblies were mostly smooth affairs. I only recall two instances where participants voted against the resolutions prepared in advance by the Executive Committee. The first time was at Manchester, where there was a question of officially recommending the digitalisation of all photographic glass plates. There was no doubt about the scientific value of such an undertaking, but it was felt that it would require huge resources; Bob Kraft, the IAU President, was quite concerned not to impose this load on all observatories. But the IAU members felt differently, and a resolution was passed to recommend the transfer of the historic observations onto modern media by digital techniques. The second time that I remember quite vividly was the discussion about the classification of Pluto at the 2006 IAU General Assembly in Prague. This was indeed a difficult moment because other solar system bodies of similar characteristics as Pluto had already been identified. The Executive Committee created a Planet Definition Committee, to recommend a resolution about this issue. The first committee did not arrive at a unified decision, so a second committee was installed that prepared a recommendation. Their recommendation was highly questioned by several astronomers, and the debates among the participants became very heated; emotions ran high. In the end, the definition of "planet" did not include Pluto. This question is still a sore point for some astronomers, complicated now with the extra-solar planets that are now being discovered at a dizzying rate.'

In the margins of the General Assemblies, the Latin American astronomers used to gather to discuss common problems among themselves.

'For some time, the IAU Officers encouraged us to organise regional meetings for Latin America. At the time, I was not very enthusiastic about it; I feared that would separate us from the mainstream astronomers in IAU. Fortunately, in 1978, the Chilean astronomers convened the first meeting in Santiago, and from thereon, the project caught on. The second meeting was in Merida, Venezuela in 1981; and by now it has become a regularly organised activity in Latin American astronomy. Clearly, my initial lack of interest was a mistake. The fourth meeting, by the way, took place in 1984 in Rio de Janeiro, where somehow unexpectedly I became involved in the organisational aspects. In retrospective, the IAU Latin American Regional Meetings, (LARIMs) have been a great success and are highly valued by the community. I should also mention, that all the proceedings of these meetings but one have been published at the Revista Mexicana de Astronomía y Astrofísica Serie de Conferencias, in which I am directly involved. I should also mention that at the Rio de Janeiro meeting, the female astronomers also began to have their own lunch gatherings under the name of *ALMA— Asociación Latino-Americana de Mujeres Astrónomas*. Later, as you know, our name was "taken away" by a larger astronomical consortium and meanwhile, these gatherings have fizzled out. The younger female astronomers do not find them necessary any more. I interpret this as progress regarding the recognition of women in our discipline.'

(Continued)

This is not to say that the issue of women's rights and gender equality was addressed up front in the IAU. In fact, it was quite the contrary. But fifteen years after the topic of women's right had gained prominence in the United States, it finally emerged also within the IAU.

'In the 1970s, the issue of women's rights had gained considerable strength in the USA. It may have done likewise in Europe, but I believe that it had not yet reached the higher echelons of academia, which were some of the dominant forces within the IAU. In fact, the Union was run by very well established male astronomers that had little interest and no inclination to deal with this topic. Although there were exceptions to this way of thinking. Thus Richard West, as an advisor to the IAU Executive Committee, encouraged me to participate in an informal gathering at the 1988 General Assembly in Baltimore to discuss the issue of women in astronomy. Vera Rubin, Cecylia Iwaniszewska, John Percy, Mazlan Othman, and I myself participated in a panel discussion. The gathering was a first step to the establishment of the Working Group on Women in Astronomy.'

Between 1982 and 1988, Manuel Peimbert, the husband of Torres-Peimbert, served as Vice-President of the IAU. Two years later, at the Manchester General Assembly, Torres-Peimbert followed.

Torres-Peimbert became IAU President at the General Assembly in Honolulu. In 2012 the IAU accepted the invitation by the US National Academy of Sciences, the U.S. National Committee for the IAU. With Hawai'i being one of the foremost centres in the world for ground-based observational astronomy, this General Assembly was thought to have been a celebration of state-of-art astronomy in a place that would also be the home of one of the largest optical telescopes ever built, the TMT. But in the meantime, the TMT had found itself at the centre of a complex conflict between activist groups in Hawai'i pursuing a range of political goals—from environmental concerns and historical grievances to religious beliefs—and the authorities that had granted permission for the TMT to be erected on Mauna Kea. For the activists, an international meeting like the IAU General Assembly was an opportunity to rally behind what they saw as a common cause, resistance to the construction of the TMT, and to air their views.

'As astronomers, we were obviously supportive of the TMT project and quite concerned about the implications of the project delay. Nonetheless, we consciously refrained from making public statements to this effect, because there was a consensus that de-escalation was the better strategy. It became a very sensitive situation and for that reason, a video address at the Opening Ceremony by President Obama, which had originally been expected, did not happen. Also, IAU as an organisation tried to reach out to the activist groups, though sadly to little avail.

Norio Kaifu, being associated with the TMT, realised how delicate this issue was; and indeed, all of us were focusing on not creating additional problems for the TMT other than those already present. We were also very concerned with the security and the normal running of the meeting itself. In the end, we spent as much time dealing with the local security officers as with our normal organisational activities.'

Thankfully, the meeting happened without any incidents. A major demonstration on 9 August gathering an estimated 10,000 Hawai'ian people was peaceful

(Continued)

and on 11 August, during a public talk about astronomy in Hawai'i, a representative of the activists had the opportunity to present their views.

'I'm quite satisfied that the presence of the General Assembly in Honolulu did not magnify the conflict or complicate an already difficult situation for our TMT colleagues.'

During her term as IAU President, much effort has gone into preparing the internal organisation for the coming years for which the IAU has set its sight high.

'Looking at the last three years, the time after the meeting in Honolulu, internally we have, once again, tightened our administrative procedures. In this respect, Piero Benvenuti has managed to reorganise the office and I find it very commendable that with the limited resources available to run the office, it is possible to run our ever-expanding activities. The organisation of the administration remains a permanent challenge, even if modern IT systems represent a considerable help.

The Office of Astronomy for Development (OAD) in collaboration with South Africa has been well established and internationally recognised, including a better understanding of the necessary level and feasibility of evaluation. The Office for Astronomy Outreach (OAO) jointly with Japan National Observatory has also found its role within the overall IAU framework. And more recently, the Office for Young Astronomers (OYA) has been formalised with the Norwegian Academy of Sciences; it supports the International Schools for Young Astronomers (ISYA), which have been running since 1967 almost on their own. They are now becoming better integrated into the overall IAU structure.

We have also developed the new Strategic Plan 2020–2030 which is much more ambitious and more encompassing, than the Strategic Plan 2010–2021; it includes the initiative to establish an Office of Astronomy for Education (OAE). This project has been prepared by a task group led by Ewine van Dishoeck. I consider that the Union is really changing its face from an organisation focussed on intramural activities within our discipline to a society that interacts with the world at large.

This change of attitudes is a direct result of the very intense activity that the International Year of Astronomy sparked. But what is important is that based on the success of that worldwide activity, we have managed to create a long-lasting—permanent, you might say—set of activities with a great potential for impact. I believe that we're an example to other professional societies and also at the level of international scientific unions, some of which are trying to follow our example.'

(Continued)

IAU President Silvia Torres-Peimbert addressing the audience at the Opening Ceremony of the GA in Vienna in 2018. Photo: C. Madsen

Torres-Peimbert ends her Presidency at an important junction in time, with a clear view towards the future.

'There are several innovations that have taken place in the last three years. We are concerned with attracting the attention of young astronomers. To this effect, we have established a yearly prize for the best PhD thesis in each of the nine Divisions, and one more for the best thesis of non-OECD countries. Other changes include the initiative to incorporate young astronomers into the Union in a special category, with a time-limited individual membership that will not impact on the financial contribution of the institutional members. I believe that these changes will stimulate the interest of the younger generation of astronomers in the IAU. In addition, we propose to create a scheme to facilitate the inclusion in IAU of new member countries of limited resources. These last two initiatives, as well as the Strategic Plan 2020–2030, will be presented to the General Assembly in Vienna; we are confident that they will be approved since they have already been presented and discussed with the National Representatives. I'm proud that we've been able to support these schemes during my time in the Presidency.

In order to have a better interaction between the Executive Committee, EC, and the Divisions, there have been joint sessions with the Division Presidents in one of the yearly EC meetings. During the last three years, we have been very lucky in that the Division Presidents have been so enthusiastic that they have tried to attend all yearly meetings, even if there was not enough travel support for these additional sessions. This interaction has been very fruitful in that they have brought fresh ideas to the organisation. In addition, each one of the six Vice-Presidents has now a specific task to monitor. In the past, in general, they had no formal responsibilities, other than their very important participation on

(Continued)

the Executive Committee decisions. I believe this change has brought out a significant improvement in sharing responsibilities. I should also point out that I have been very lucky that Piero Benvenuti has been the General Secretary. We should recognise that as the IAU activities have grown in complexity, the workload on the General Secretary has been extremely demanding, so it is mandatory to have a very capable colleague to take up this responsibility.'

In 2011, Torres-Peimbert has received the l'Oréal-UNESCO Award for Women in Science in 2011 for her work determining the chemical composition of nebulae. In the following year, she was awarded the Hans A. Bethe Prize for her work in determining the quantities of helium and other elements during the development of the universe.

'Until now, three female astronomers—and IAU members—have received the l'Oreal-UNESCO Prize. Yet there's no direct connection between the l'Oreal Prize and the IAU. But it's noteworthy that in Latin America, several of the outstanding women in the physical sciences are in astronomy. Originally the l'Oreal-UNESCO Prize was awarded every two years only in the area of Material Science, and now this definition has been enlarged to include Physical, Mathematics and Computing Sciences.'

Despite her personal success, like so many other female scientists of her generation, Torres-Peimbert has experienced the uphill struggle because of her gender. But times are changing and with them many attitudes. Torres-Peimbert, the second female IAU President, will be succeeded by Ewine van Dishoeck who in turn will be supported by another prominent astronomer, Teresa Lago, as IAU General Secretary. If the struggle was long and cumbersome, it has paid off— also in the top echelons of the IAU.

'I'm delighted that Ewine van Dishoeck will be the following President, and that Teresa Lago the new General Secretary. Both are very dynamic and full of ideas. Ewine has been very active in several very valuable initiatives, and Teresa has already been deeply involved in the organisation of the Symposia and the General Assembly itself. I believe that IAU can look forward to a bright future. Not a bad perspective for an organisation that will soon celebrate its 100th birthday.'

Teresa Lago, IAU Secretary General at a Unique Moment in the History of the Union

Assistant General Secretary (2015–2018), General Secretary (2018–2021)
    Interview conducted in Vienna on 24 August 2018
    The team taking the IAU into its second century is led by the third female president of the IAU and the third female general secretary. A president with a long IAU association and a general secretary, hugely experienced, but not tied to IAU traditions opens new perspectives for the Union.

(Continued)

Teresa Lago in the AGS Office at the IAU General Assembly in Vienna. Photo: C. Madsen

Teresa Lago was born in Lisbon in 1947, but soon moved with her family to Africa.

'I really became an astronomer by coincidence. I was born in Lisbon but spent my childhood and youth in Angola, before taking up university studies in mathematics at the University of Porto. During my third year at Porto, I had to take an astronomy course. I realised that there was a field where one could make real discoveries. I subsequently switched from math to surveing engineering, because that included some astronomical geodesy. Then I attended a summer course at the Royal Greenwich Observatory (RGO) in Herstmonceux Castle in the UK. By then, I was completely hooked.

Returning to Portugal in 1979 was challenging. There was no research in astronomy and no active astronomers, but I did get a job in the Applied Mathematics Department. I was allowed to visit the RGO to work on data—in the night, when the astronomers there did not use the facilities. But I had to pay my way myself. Also this was the time, when the International Ultraviolet Explorer was launched. My thesis was on magnetically driven winds of low mass stars and IUE observations were very interesting to me. Also, travelling to Madrid to observe with IUE I could go by bus. Again, I had to pay for this myself, since there was no funding available. But there was no-one in Portugal that I could work with. To alleviate the problem, I managed to negotiate a collaboration between the Applied Mathematics Department and the Physics Department of the University to open a new undergraduate course for astronomy with applied math and physics. We started out with 20 positions, but had no less than 200 applicants.'

In 1986 José Mariano Gago, arguably one of Europe's most prolific science policymakers, became president of the Portuguese Research Council. That not only opened the door to research funding, but he also asked Lago to develop an astronomy programme for Portugal. Among other things this led to the establishment of the Centre for de Astrophysics at the University of Porto (CAUP) that Lago led for eighteen years and also to Portugal joining ESO, including a 10-year transition period during which Portuguese astronomy could develop.

(Continued)

Both Gago and Lago were strong advocates of scientific outreach to stimulate public awareness and interest in science and technology.

'When I started the Centre for Astrophysics in Porto (CAUP), I included in the Centre regulations the requirement that all staff should devote 10% of their time to do science outreach. We were probably the first institute in Europe to do that. This came in quite handy when, in 1996, we established the Porto Planetarium run by CAUP in a collaboration between the University, the Municipality of Porto and the Portuguese Ministry for Science and Technology.'

In the early years of the new century, Lago became involved in the preparations and setting up of the European Research Council (ERC). This was a seminal step for EU research funding. Today, the European Research Council runs a highly coveted grant scheme embedded in the EU Framework Programmes for funding of science and technology, but with scientific excellence as the only criterion for funding.

'We had a lot of discussions and also quite a bit of a fight, but we gained a lot that had never been done before in this system: The basis was "trust the researchers", which was foreign to EU administrators. And as the fund was set up with scientific panels to assess the grant applications, I argued—successfully—for the establishment of a dedicated panel for astronomy.

When I returned to Portugal in 1979, I was keen to join the IAU. But it was not trivial, because while we had a national representative, we had no astronomers. Because of my interest in teaching, I also wanted to join Commission 46, but I was told that Portugal already was represented in that Commission, so it was 'not necessary' that I'd become a member! Eventually though, I did make it! Once more, as for a long time I had no travel funds, I could not afford to travel to General Assemblies, but I did attend the IAU Regional Meeting in Prague in 1987.'

With her background, Teresa Lago is not steeped in the ways and traditions of the IAU.

'I am not tied to the traditions. Traditions are important and I have respect for them, but I am the one to ask questions and ready to help the IAU to move on. I believe that the IAU can do a lot in terms of inclusion in all aspects and meanings of the word. For sure, the IAU has already done a lot, but we can do even more!

I left the ERC in 2013 and since I had also asked for early retirement from the University, I had time to do a long list of discovery projects and I'm still curious, so when I was approached by Thierry Montmerle about working for the IAU, I accepted the challenge. As Assistant General Secretary, I have been in charge of the symposia, the focus meetings and the scientific programme for the General Assembly in Vienna. I've also been an ex-officio member of the Steering Committee for the OAD and also acted as the contact person for the OAO, which has been very interesting. I have been to the Paris Office several times to work with Piero Benvenuti and the office is now working very well.

I am very happy that now we have the first full and well balanced strategic programme addressing all the interests and obligations of the Union. It's a manifestation of an organisation that is growing and active. Will it be challenging? Well, a dead structure is not difficult to manage, but we don't want a dead structure, we want a lively and creative structure. One example is the special membership category for young astronomers. For them we have created a junior members' working group under the Executive Committee and we want it to be creative and innovative. So, at 100 years, the IAU is rejuvenating. Our science is as vibrant as ever and so is the IAU. It's a special moment, I think, and I'm really thrilled about it.'

(Continued)

Teresa Lago clearly brings to the IAU table considerable experience, having been a driver for developing astronomy in her home country, having been a member of numerous advisory panels and committees in other countries and having worked at a high level in European science policy. She also made her organisational and managerial skills available in other contexts, including a three-year stint as President for 'Porto 2001—European Capital of Culture'.

'I accepted this task on the condition that we could include science outreach in the activities of the European Capital of Culture.'

Astronomy both as a science and as a source of inspiration for society is the theme that has determined her professional life to this day and will do so during her term as IAU General Secretary.

Ewine van Dishoeck: Presiding over the IAU@100 and Looking into the Future of This Venerable Union

President of Division H Interstellar Matter and Local Universe (2012–2015), President-Elect (2015–2018), President (2018–2021)

Interview conducted in Leiden on 19 April 2017

'The IAU has been and still is the way to get the world-wide community of astronomers organised. It is also a face for that discipline towards the outside world. It's a powerful face with content, not just in terms of science but also showing what we bring to the rest of society.'

Ewine van Dishoeck at her office in Leiden. Photo: C. Madsen

These words by Ewine van Dishoeck, IAU President Elect (2015–2018) and IAU President (2018–2021) encapsulate her view on the IAU and its role in the time to come. Not just forward-looking, it is also based on her long association with the IAU in various roles and functions.

'My first encounter with the IAU was in 1979, at the XVIIth General Assembly in Montreal. Together with my husband—we were still master students at the time—we were invited to participate as guests by Adriaan Blaauw, who was the President of the Union. Together with other people, we camped during the

(Continued)

meeting. I was still in chemistry, and this was my first opportunity to attend a big, international conference. It was exciting. Many legends were at the General Assembly—I remember seeing Chandrasekhar. Also, Blaauw told us in the corridors about the effort to solve the China/Taiwan issue, thus exposing us a little bit of the diplomacy that also goes on in the IAU.

My next General Assembly was the one in 1985 in New Delhi. It was followed by the first IAU Symposium on Astrochemistry, Symposium 120, organised by my mentor Alexander Dalgarno in Goa. This involved about 100 people from all over the world and 25 of us took a five-day tour from New Delhi to Goa. It brought this international community of people together in a way that I had never seen before—the trip really cemented people together. In 1991 I became first the secretary and then later the chair of the IAU working group on astrochemistry and so, for 20 years, I organised these IAU symposia—in the Netherlands, Korea, the USA and Spain. The latest one took place in Chile in 2017, highlighting exciting ALMA results. I think that it can honestly be said that the IAU symposia played an incredible role in shaping the field of astrochemistry. There are other cases, such as planetary nebulae. It is perhaps less visible in other fields, but Astrochemistry is a good example of how the IAU, thanks to early leadership, has really been very important. Through these activities, it became natural for me to be open to further involvement in the IAU.

In 2012 Bob Williams asked me to become President of Division H. Normally division presidencies are decided by a ballot, but this was an unusual situation because the divisions were being restructured so that ballots could not be carried out. Aside from the task itself, it was important for me to see what happens in the Executive Committee, whose meetings Division Presidents were invited to attend.

Accepting a functional duty like this always raises the question of balancing it with one's scientific career—a question with which most senior university-based scientists staff are familiar. For me, however, the IAU constituted an interesting and important "scenery" as compared to other functions that senior academic staff normally must accept, and so, accepting this honour and challenge was not difficult.'

Clearly, taking upon herself these tasks for the IAU has in no way dented van Dishoeck's sterling scientific career, as her impressive CV bears witness to. After obtaining her PhD from Leiden University in 1984, she has, among other things, served as Visiting Professor at Princeton University, Assistant Professor of Cosmochemistry at Caltech, Professor of Molecular Astrophysics, University of Leiden and Scientific Director of NOVA, the Netherlands Research School for Astronomy. She has received several prestigious medals and awards (including the 2018 Kavli Prize), is co-Editor of the Annual Review of Astronomy and Astrophysics and has been involved in a range of major observational facility projects, including not the least the Atacama Large Millimeter/submillimeter Array in Chile.

Like for many of her female colleagues with impressive scientific achievements, what used to be the exclusive realm of 'men of science' had eventually to change. Although she will not place too much emphasis on it herself, van Dishoeck's role in the history of the IAU is also an expression of a bigger change, albeit a slow one, within the Union. In the course of its 100-year long history, as of 2018, only 14 female astronomers have made it to the IAU Executive Committee, including three General Secretaries and three Presidents. However, half of them have served within the last decade.

(Continued)

'The IAU has of course been very male oriented. Wilhelmina Iwanowska was the first Vice-President, back in 1973, but since then there has not been that many until more recently.

When he set up the new divisions Bob Williams did a lot to remedy the situation by picking new Division Presidents, with more than half of them being female—also to show that women could take a leading role in these activities. Also, the Executive Committee is now at a 50/50 level, so there is a now a better balance there. However, overall, the statistics are still horrendous. In 1987, female membership was below 10% and even today, female scientists constitute only 16% which is also due to the fact that only well-established scientists are proposed for membership. This is at least partly an effect of appointment practices by the universities and institutes, over which the IAU has no control. But we can bring the numbers into the open and challenge countries to do better. We also encourage the active participation of female astronomers in the IAU and its activities.'

The IAU centenary, of course, offers an opportunity to reflect on the past but also to ponder about the time to come. As President of the IAU at this juncture, what does she as important tasks for the IAU in the years to come?

'Among the traditional tasks of the IAU we have seen the setting of astronomical standards, naming of celestial objects and, of course, the organisation of scientific symposia and colloquia. However, there has been an explosion in scientific meetings. So, while the IAU is still a guarantor for maintaining a truly global aspect of scientific meetings, overall it is a less significant activity than it perhaps used to be. For that reason, important changes were introduced in the past decade, triggered not the least by George Miley. The cover of the Strategic Plan, adopted in 2009, provides a strong illustration of the many areas in which astronomy has a role to play, including the technology part, the science part and the multi-disciplinary aspect, and the culture and society part. The plan itself has given the IAU a much-expanded scope than it had in the past. This also pertains to our geographical reach. The establishment of the Regional OAD offices, as well the introduction of IAU regional meetings some 20 years ago, has strengthened the global importance of the IAU. But there's more to be done:

We need to make our face to the world more explicit, to show that astronomy is an incredibly exciting topic that is not just important for astronomy's own sake but is important for stimulating young people's interest in the sciences in general. But also to realise the growing necessity of a multi-disciplinary approach to some of the questions that astronomers ask. If, for example, we wish to address the question: "Are we alone in the Universe?" then we need geologists, biologists, etc. to come and work with us.

And we have the high-tech aspect of astronomy. We're building fantastic new facilities, and for that, we need the best engineers and the brightest minds also in the future. Historically, astronomy has been an important driver for new technologies and will continue to be so, but this is not always fully understood by the decision-makers or indeed by society at large.

That's why I think that the makers of the cover illustration of the Strategic Plan had all the right ingredients and our centenary provides us with a unique chance demonstrate the importance of astronomy to the world, not the least by building on the fantastic legacy of the International Year of Astronomy.'

The Present and Future of the IAU

Our understanding of the Universe in which we live has changed dramatically since the founding of the IAU, just one century ago. The Universe is larger, more diverse and more dynamic than anybody then could have imagined. IAU members and officials have played their role in this evolution of our knowledge, as researchers as well as facilitators of the international exchange of ideas.

Much progress has been made. In 2018, President-Elect Ewine van Dishoeck was awarded the prestigious Kavli Prize in Astrophysics, '*for her combined contributions to observational, theoretical, and laboratory astrochemistry, elucidating the life cycle of interstellar clouds and the formation of stars and planets*'.[23] At the last General Assembly, major topics of discussion and/or Invited Discourses included gravitational waves, exoplanets, and the results of the Gaia mission—and plans for many future facilities.

Much is still unknown, however. The science of exoplanets has proved to offer a large new realm to explore, and the field is still young. Exoplanet imaging is on the top of the target list for the planned large telescopes. The main prizes are finding Earth-like planets, and detecting signs of biological activity in exoplanetary atmospheres—anything to get closer to understanding life in the Universe.

And while our cosmological models seem to be robust and able to accurately explain many observations with a handful of known parameters, the very nature of dark matter and dark energy still remains a mystery. Nobody knows where the next breakthrough is to be expected, but it is possible that the very foundations of physics have to be revised.

In the meantime, a new generation of telescopes—also optical—will undoubtedly provide new and exciting observations. Every time a new, bigger, or better instrument is pointed at the sky, something unexpected turns up. We can now observe the entire electromagnetic spectrum as well as other 'messengers', such as neutrinos and other cosmic particles. The recent detection of gravitational waves has just opened another new window. Who knows what we will see through it?

Closer to home, astronomers will keep looking for potentially hazardous Near Earth Objects—one of the very few natural dangers confronting the Earth, which can be predicted and potentially averted, provided that a warning is issued long enough in advance. Actual mitigation measures are beyond

---

[23] Kavli Prize in Astrophysics citation 2018, http://kavliprize.org/

the realm of astronomy (and certainly beyond the means of the IAU), but the search for and scientific study of NEOs, and the accurate determination and computation of their orbits, are clearly jobs for astronomers. Thus, the IAU and its Minor Planet Center are playing key roles in this effort, which is truly of global importance. Time will tell if and when it will be needed, but at least preparations are now beginning to be made.

To facilitate and promote the study of our fascinating universe is the key assignment of the IAU. The way in which it has done this, has changed however, just as the science of astronomy has changed. In its first century, the IAU has made the transition from a rather closed academic society to a Union with an active social role and with a strong public profile. Today, it focuses on using astronomy as a natural gateway into the natural sciences, to improve scientific literacy as well as by showing the beauty of our quest to understand the Universe we live in. The appetite for the kind of comprehensive 'capacity building' programme described in the new *Strategic Plan* appears insatiable, and the response so far is overwhelmingly positive. For society at large, this may well be the most important—and most enduring—legacy of the IAU's first century.

During the first half of its existence, the role of the IAU within the astronomical community was also established. Despite the initial political controversies about German membership, the authority of the IAU as the central coordinator of nomenclature and terminology was generally accepted, and the IAU General Assemblies became firmly anchored as the main international conferences in the field. The large number of active Commissions functioned as central fora for specific themes. They also made it possible for a relatively large number of researchers to get actively involved in the Union—many more than just the representatives of the member countries. This developed into individual membership, one of the unique features that have made the IAU the most lively and active of scientific unions. Some other unions have started individual membership programmes, but they have nowhere near the reach of the IAU.[24]

After the Second World War, the IAU gradually expanded its activities: sponsoring scientific meetings (notably Symposia), supporting individual researchers through exchange programmes, and stimulating educational and outreach projects. But its main role remained to provide a neutral and impartial platform for international scientific meetings, even in times of geopolitical conflict.

---

[24] For example the IUPAC Affiliate Membership Program: https://iupac.org/home/individual-members/

Today, the IAU aims to include the entire community, including those who are just starting their careers. The total membership count (including junior members) is now more than 13,600, of which more than 11,000 are registered as 'active members'.[25] The criterion for full membership is now 'a PhD related to astronomy', although the national organisations that nominate members sometimes include extra rules—exercising independent judgement and making exceptions has always been a hallmark of the IAU. The number of member countries is expected to keep rising, aided by the various development programmes.

In 2018, a new, comprehensive Strategic Plan was approved by the General Assembly in Vienna, in which the IAU defines yet more ambitious aims for education and development. The new plan is much broader than its predecessor, covering all aspects of the IAU's activities, including plans for a new Office of Astronomy for Education. Education and development will undoubtedly serve as one of the foundations of the life and benefits of the IAU in its second century.

The new Strategic Plan also aims to give the IAU a role in building a science-based consensus in preparation for the next generation of new global-scale infrastructures. This has been a recurrent ambition of the Union. It is now more urgent than ever, since the instruments have become so large and expensive, and the number of possible sites for them so limited, that only a single one in each class can be realised at a time—in global collaboration and by coordination of its use.

At the foundation of the IAU in 1919—now exactly a century ago—one of the two aims of the Union was stated to be *to promote the study of astronomy in all its departments*. In the early twenty-first century, this aim took on a whole new dimension. In the new Strategic Plan 2020–2030, the goals of the Union are re-formulated as follows:

Goal 1: the IAU leads the worldwide coordination of astronomy and the fostering of communication and dissemination of astronomical knowledge among professional astronomers

Goal 2: the IAU promotes the inclusive advancement of the field of astronomy in every country

Goal 3: the IAU promotes the use of astronomy as a tool for development in every country

---

[25] The latest membership count, including some statistics, can be found at: https://www.iau.org/administration/membership/

Goal 4: the IAU engages the public in astronomy through access to astronomical information and communication of the science of astronomy

Goal 5: the IAU stimulates the use of astronomy for teaching and education at school level

The IAU now emphatically presents itself as *the* global promoter of astronomy, both within the scientific community and among the wider public. Its audience is no longer just astronomers, but the entire world population, from policy makers to school-age children. This marks a fundamental change in the way the IAU is serving the astronomical community.

You have the book in your hands and can form your own opinion—both of the book itself, and the story it tells of the metamorphosis of the IAU. Now it has a clear blueprint for its next century.

The IAU Strategic Plan 2020–2030

Appendix A: General Assemblies, Membership Numbers, Officials

The membership and participant numbers up to 1985 are based on numbers published in IB 56 pp. 6–7 and the GA newspaper from New Delhi, 1985. These can differ slightly from the numbers provided by Blaauw (1994).

1919 Foundational meeting, Brussels, Belgium

President: Benjamin Baillaud (France)
General Secretary: Alfred Fowler (United Kingdom)

I 1922 Rome, Italy

No. of individual members: 207
No. of GA participants: 83
Incoming President: William W. Campbell (USA)
Incoming General Secretary: Alfred Fowler (United Kingdom)

II 1925 Cambridge, England, United Kingdom

No. of individual members: 244
No. of GA participants: 189
Incoming President: Willem de Sitter (Netherlands)
Incoming General Secretary: Frederick J.M. Stratton (United Kingdom)

© Springer Nature Switzerland AG 2019
J. Andersen et al., *The International Astronomical Union*,
https://doi.org/10.1007/978-3-319-96965-7

III 1928 Leiden, Netherlands

No. of individual members: 288
No. of GA participants: 261
Incoming President: Frank W. Dyson (United Kingdom)
General Secretary: Frederick J.M. Stratton (United Kingdom)

V 1932 Cambridge, Massachusetts, United States

No. of individual members: 406
No. of GA participants: 203
Incoming President: Frank Schlesinger (USA)
General Secretary: Frederick J.M. Stratton (United Kingdom)

V 1935 Paris, France

No. of individual members: 496
No. of GA participants: 317
Incoming President: Ernest Esclangon (France)
General Secretary: Frederick J.M. Stratton (United Kingdom)

VI 1938 Stockholm, Sweden

No. of individual members: 554
No. of GA participants: 293
President: Arthur S. Eddington (United Kingdom) until 1944, then Harold
  Spencer-Jones (United Kingdom)
General Secretary: Jan H. Oort (Netherlands); from 1940-45 acting secretary:
  Walter S. Adams (USA)

1946 Copenhagen Conference, Denmark

President: H. Spencer-Jones (United Kingdom)
General Secretary: Jan H. Oort (Netherlands)

VII 1948 Zürich, Switzerland

No. of individual members: 611
No. of GA participants: 279
Incoming President: Bertil Lindblad (Sweden)
Incoming General Secretary: Bengt Strömgren (Denmark; resigned 1951)

VIII 1952 Rome, Italy

No. of individual members: 809
No. of GA participants: 434
Incoming President: Otto Struve (USA)
Incoming General Secretary: Pieter T. Oosterhoff (Netherlands)

IX 1955 Dublin, Republic of Ireland

No. of individual members: 888
No. of GA participants: 597
Incoming President: André Danjon (France)
General Secretary: Pieter T. Oosterhoff (Netherlands)

X 1958 Moscow, Soviet Union

No. of individual members: 1127
No. of GA participants: 820
Incoming President: Jan H. Oort (Netherlands)
Incoming General Secretary: Donald H. Sadler (United Kingdom)

XI 1961 Berkeley, California, United States

No. of individual members: 1289
No. of GA participants: 765
Incoming President: Viktor A. Ambartsumian (Soviet Union)
General Secretary: Donald H. Sadler (United Kingdom)

XII 1964 Hamburg, West Germany

No. of individual members: 1630
No. of GA participants: 1160
Incoming President: Pol Swings (Belgium)
Incoming General Secretary: Jean-Claude Pecker (France)

XIII 1967 Prague, Czechoslovakia

No. of individual members: 2009
No. of GA participants: 1835
Incoming President: Otto Heckmann (Germany)
Incoming General Secretary: Luboš Perek (Czechoslovakia)

XIV 1970 Brighton, England, United Kingdom

No. of individual members: 2590
No. of GA participants: 2255
Incoming President: Bengt Strömgren (Denmark)
Incoming General Secretary: Cornelis de Jager (Netherlands)

XV 1973 Sydney, New South Wales, Australia

No. of individual members: 3188
No. of GA participants: 840
Incoming President: Leo Goldberg (USA)
Incoming General Secretary: Georgios Contopoulos (Greece)

1973 extraordinary GA in Torún, Poland

No. of GA participants: 813

XVI 1976 Grenoble, France

No. of individual members: 3805
No. of GA participants: 2134
Incoming President: Adriaan Blaauw (Netherlands)
Incoming General Secretary: Edith A. Müller (Switzerland)

XVII 1979 Montreal, Quebec, Canada

No. of individual members: 4504
No. of GA participants: 1965
Incoming President: M.K. Vainu Bappu (India)
Incoming General Secretary: Patrick A. Wayman (Ireland)

XVIII 1982 Patras, Greece

No. of individual members: 5200
No. of GA participants: 1700
Incoming President: Robert Hanbury-Brown (United Kingdom)
Incoming General Secretary: Richard M. West (Denmark)

XIX 1985 Delhi, India

No. of individual members: 6025
No. of GA participants: 1400
Incoming President: Jorge Sahade (Argentina)
Incoming General Secretary: Jean-Pierre Swings (Belgium)

XX 1988 Baltimore, Maryland, United States

No. of individual members: 6711
No. of GA participants: 1990
Incoming President: Yoshihide Kozai (Japan)
Incoming General Secretary: Derek McNally (United Kingdom)

XXI 1991 Buenos Aires, Argentina

No. of individual members: 7301
No. of GA participants: 1148
Incoming President: Alexander Boyarchuk (Russia)
Incoming General Secretary: Jacqueline Bergeron (France)

XXII 1994 The Hague, Netherlands

No. of individual members: 8328
No. of GA participants: 1700
Incoming President: Lodewijk Woltjer (Netherlands)
Incoming General Secretary: Immo Appenzeller (Germany)

XXIII 1997 Kyoto, Kansai, Japan.

No. of individual members: 8562
No. of GA participants: 2100
Incoming President: Robert P. Kraft (USA)
Incoming General Secretary: Johannes Andersen (Denmark)

XXIV 2000 Manchester, England, United Kingdom

No. of individual members: 8737
No. of GA participants: 1550
Incoming President: Franco Pacini (Italy)
Incoming General Secretary: Hans Rickman (Sweden)

XXV 2003 Sydney, New South Wales, Australia

No. of individual members: 9114
No. of GA participants: 2050
Incoming President: Ronald D. Ekers (Australia)
Incoming General Secretary: Oddbjørn Engvold (Norway)

XXVI 2006 Prague, Czech Republic

No. of individual members: 9783
No. of GA participants: 2412
Incoming President: Catherine J. Cesarsky (France)
Incoming General Secretary: Karel van der Hucht (Netherlands)

XXVII 2009 Rio de Janeiro, Brazil

No. of individual members: 10,144
No. of GA participants: 2434
Incoming President: Robert Williams (USA)
Incoming General Secretary: Ian Corbett (United Kingdom)

XXVIII 2012 Beijing, China

No. of individual members: 10,904
No. of GA participants: 2710
Incoming President: Norio Kaifu (Japan)
Incoming General Secretary: Thierry Montmerle (France)

XXIX 2015 Honolulu, Hawai'i, United States

No. of individual members: 12,376
No. of GA participants: 2581)
Incoming President: Sylvia Torres-Peimbert (Mexico)
Incoming General Secretary: Piero Benvenuti (Italy)

XXX 2018 Vienna, Austria

No. of individual members: 13.595, including 352 Junior Members
No. of GA participants: 3004
Incoming President: Ewine F. van Dishoeck (Netherlands)
Incoming General Secretary: M. Teresa V.T. Lago (Portugal)

XXXI 2021 Busan, South Korea

Incoming President: Debra M. Elmegreen (USA)
Incoming General Secretary: Ian Robson (United Kingdom)

XXXII 2024 Cape Town, South Africa

Appendix B: Commission History

Commissions (1922–2015)

Noting merging, termination or change of name. Some commission numbers were re-used.

NB Originally, the official names of the Commissions were only in French; from 1958 the names were also listed in English. Later years without comment: name updated or number re-used.

Based on the Transactions of the General Assemblies, see https://www.iau. org/publications/iau/transactions_b/

The list deviates a bit from the lists provided at https://www.iau.org/science/scientific_bodies/past_commissions/

1. 1922 Relativity
   1925 Terminated
2. 1922 Ancient Manuscripts (never approved)
3. 1922 Notations and Units and Economy of Publication
   1958 Notations
   1961 Terminated
4. 1922 Ephemerides
5. 1922 Abstracts and Bibliography
   1937 Documentation
   1979 Documentation and Astronomical Data
6. 1922 Astronomical telegrams

© Springer Nature Switzerland AG 2019
J. Andersen et al., *The International Astronomical Union*,
https://doi.org/10.1007/978-3-319-96965-7

7. 1922 Dynamical Astronomy and Astronomical Tables
   1932 Terminated
   1948 Celestial Mechanics
   1997 Celestial Mechanics and Dynamical Astronomy
8. 1922 Meridian Astronomy
   1958 Positional astronomy
   2000 Astrometry
9. 1922 Optical Research
   1928 Astronomical Instruments
   1985 Instruments and Techniques
   2000 Instrumentation and Techniques
   2006 Terminated
10. 1922 Solar Radiation
    1925 Terminated
    1932 Sunspots and Characteristic Figures of the Sun
    1948 Photospheric Phenomena
    1958 Solar Activity
11. 1922 Spectro-register of Velocities (never approved; topic merged with C12)
    1932 Chromospheric Phenomena
    1948 Outer Layers of the Sun
    1958 Merged with C12
12. 1922 Solar Atmosphere
    1925 Solar Physics
    1932 Solar Radiation and Solar Spectroscopy
    1958 Radiation and Structure of the Solar Atmosphere (merged with C11 and C13)
    1997 Solar Radiation and Structure
13. 1922 Astronomical expeditions (never approved)
    1932 Solar Eclipses
    1958 Merged with C12
14. 1922 Wave-Length Standards and Tables of Spectra
    1961 Fundamental Spectroscopic Data
    1979 Atomic and Molecular Data
15. 1922 Solar Rotation
    1928 Terminated
    1935 Physical study of Comets
    1973 Physical study of Comets, Minor Planets and Meteorites
    2000 Physical study of Comets and Minor Planets
16. 1922 Physical Observations of the Planets

1925 Physical Observations of the Planets, Comets and Satellites
1935 Physical Observations of the Planets and Satellites
1955 Physical Study of Planets and Satellites.

17. 1922 Lunar Nomenclature
1938 The Moon
1948 Motion and Figure of the Moon
1964 The Moon
1979 Terminated

18. 1922 Longitudes by Wireless Telegraph
1952 Geographical positions
1958 Terminated

19. 1922 Variation of Latitude
1964 Rotation of the Earth
(1988 Subject transferred to IERS, but the Commission was not terminated)

20. 1922 Minor Planets
1925 Positions and Ephemerides of Minor Planets, Comets and Satellites
1958 Positions and Motions of Minor Planets, Comets and Satellites

21. 1922 Comets
1958 Light of the Night Sky
2009 Galactic & Extragalactic Background Radiation

22. 1922 Shooting Stars
1935 Shooting Stars, Zodiacal Light and Similar Problems
1955 Meteors and Meteorites
1976 Meteors and Interplanetary Dust
2000 Meteors, Meteorites and Interplanetary Dust

23. 1922 Carte du Ciel
1970 Merged with C24

24. 1922 Stellar Parallaxes
1928 Stellar Parallaxes and Proper Motions
1970 Photographic Astrometry (merged with C23)
2000 Terminated (merged with C8)

25. 1922 Stellar Photometry
1973 Stellar Photometry and Polarimetry
2012 Astronomical Photometry and Polarimetry

26. 1922 Double Stars
1994 Double and Multiple Stars

27. 1922 Variable Stars

28. 1922 Nebulae
1928 Nebulae and Star Clusters

1948 Extra-Galactic Nebulae
1961 Galaxies

29. 1922 Spectral Classification of Stars
    1928 Stellar Spectra

30. 1922 Stellar Radial Velocities
    1964 Radial velocities

31. 1922 Time (including Bureau Internationale de l'Heure)
    (1988 Subject transferred to IERS, but the Commission was not terminated)

32. 1922 Calendar Reform (immediately terminated)
    1932 Selected Areas
    1958 Terminated (merged with C33)

33. 1928 Stellar Statistics
    1955 Structure and Dynamics of the Galactic system

34. 1925 Solar Parallax
    1938 Interstellar Matter
    1948 Interstellar Matter and Galactic Nebulae
    1958 Interstellar Matter and Planetary Nebulae
    1982 Interstellar Matter

35. 1928 Stellar constitution

36. 1935 Spectrophotometry
    1961 Theory of Stellar Atmospheres

37. 1948 Star Clusters
    1955 Star Clusters and Associations

38. 1948 Exchange of Astronomers
    2000 Terminated (merged with C46)

39. 1948 International Observatories
    1955 Terminated

40. 1948 Radio Observations
    1952 Radio astronomy

41. 1948 History of astronomy

42. 1948 Photometric Double Stars
    1973 Close Binary Stars

43. 1958 Magneto-Hydrodynamics and Physics of the Ionised Gases
    1967 Astrophysical Plasmas and Magnetohydrodynamics
    1973 Terminated

44. 1958 Astronomical Observations from Outside the Terrestrial Atmosphere
    1979 Observations from Space
    1994 Space and High Energy Astrophysics (merged with C48)

45. 1964 Spectral Classifications and Multi-Band Colour Indices

1979 Stellar Classification
46. 1964 Teaching of Astronomy
2000 Astronomy Education and Development (merged with C38)
47. 1970 Cosmology
48. 1970 High Energy Astrophysics
1994 Merged with C44
49. 1973 Interplanetary Plasma & Heliosphere
50. 1973 Protection of Existing & Potential Observatory Sites
51. 1982 Bio-Astronomy: Search for Extraterrestrial Life
1991 Bio-Astronomy
52. 2006 Relativity in Fundamental Astronomy
53. 2006 Extrasolar Planets
54. 2006 Optical & Infrared Interferometry
55. 2006 Communicating Astronomy with the Public

1973 Working Group for Planetary System Nomenclature

Divisions (1994–2012)

I. Fundamental Astronomy
II. Sun & Heliosphere
III. Planetary Systems Sciences
IV. Stars
V. Variable Stars
VI. Interstellar Matter
VII. Galactic System
VIII. Galaxies & the Universe
IX. Optical & Infrared Techniques
X. Radio Astronomy
XI. Space & High Energy Astrophysics

Added in 2003:

XII. Union-Wide Activities

New Divisions (2012) and Commissions (2015)

A Fundamental Astronomy

A1 Astrometry
A2 Rotation of the Earth
A3 Fundamental Standards
A4 Celestial Mechanics and Dynamical Astronomy, *joint with Division F*

B Facilities, Technologies and Data Science

B1 Computational Astrophysics
B2 Data and Documentation
B3 Astroinformatics and Astrostatistics
B4 Radio Astronomy
B5 Laboratory Astrophysics
B6 Astronomical Photometry and Polarimetry
B7 Protection of Existing and Potential Observatory Sites, *joint with Division C*

C Education, Outreach and Heritage

C1 Astronomy Education and Development
C2 Communicating Astronomy with the Public
C3 History of Astronomy
C4 World Heritage and Astronomy

D High Energy Phenomena and Fundamental Physics

D1 Gravitational Wave Astrophysics

E Sun and Heliosphere

E1 Solar Radiation and Structure
E2 Solar Activity
E3 Solar Impact Throughout the Heliosphere

F Planetary Systems and Bioastronomy

F1 Meteors, Meteorites and Interplanetary Dust
F2 Exoplanets and the Solar system
F3 Astrobiology

G Stars and Stellar Physics

G1 Binary and Multiple Star Systems
G2 Massive Stars
G3 Stellar Evolution
G4 Pulsating Stars
G5 Stellar and Planetary Atmospheres

H Interstellar Matter and Local Universe

H1 The Local Universe
H2 Astrochemistry
H3 Planetary Nebulae
H4 Stellar Clusters throughout Cosmic Space and Time, *joint with Divisions G and J*

J Galaxies and Cosmology

J1 Galaxy Spectral Energy Distributions, *joint with Divisions D, G, and H*
J2 Intergalactic Medium, *joint with Division B and H*

Cross-Division Commissions

X1 Supermassive Black Holes, Feedback and Galaxy Evolution, *Divisions D and J*
X2 Solar System Ephemerides, *Divisions A and F*

Appendix C: List of National Members

By date of admission

This list is based on https://www.iau.org/administration/membership/national/, with corrections and additions based on the IAU Transactions B: Proceedings of the General Assemblies.

Some countries are listed twice; usually, this is because they resigned or were suspended earlier (mostly for financial reasons). The records are not always clear, however. In some cases, it just the adhering body changed; e.g., in South Africa, or in Australia in 1938–1939. Countries marked * are not currently members.

Changes in the names of countries are not listed, but if countries broke up, parts that joined separately are listed as new members.

1920	Belgium
1920	Canada
1920	France
1920	Greece
1920	Japan
1920	United Kingdom
1920	United States of America
1921	Italy
1921	Mexico
1922	Australia
1922	Brazil
1922	Czecho-Slovakia*
1922	Denmark

© Springer Nature Switzerland AG 2019
J. Andersen et al., *The International Astronomical Union*,
https://doi.org/10.1007/978-3-319-96965-7

1922	Netherlands
1922	Norway
1922	Poland
1922	Romania
1922	South Africa
1922	Spain
1923	Switzerland
1924	Portugal
1925	Egypt
1925	Sweden
1927	Argentina
1932	Vatican State
1935	China
1935	USSR*
1935	Yugoslavia*
1938	South Africa (rejoining)
1939	Australia (rejoining)
1946	India
1947	Hungary
1947	Chile
1947	Ireland
1948	Finland
1951	Germany
[1952	Japan re-admitted]
1953	Venezuela
1954	Israel
1954	Lebanese Republic
1955	Austria
1957	Bulgaria
1959	Taiwan
1961	Turkey
1961	Brazil (rejoining)
1961	DPR Korea
1964	Germany split: BRD & GDR
1964	Indonesia
1964	New Zealand
1967	Colombia
1969	Iran
1970	Uruguay
1970	Cuba
1973	Republic of Korea
1967	Iraq
1979	Indonesia (rejoining)
1982	China—Nanjing & Taipei

1988	Algeria
1988	Iceland
1988	Malaysia
1988	Morocco
1988	Peru
1988	Saudi Arabia
1990	Germany re-unified
1992	Russian Federation
1992	Estonia
1993	Czech Republic
1993	Slovakia
1993	Lithuania
1993	Tajikistan
1993	Ukraine
1994	Armenia
1994	Georgia
1994	Croatia
1996	Latvia
1997	Azerbaijan*
1997	Central American Assembly of Astronomers (CAAA: Costa Rica, El Salvador, Guatemala, Nicaragua, Panama)*
1998	Bolivia
1998	Macedonia, Former Yugoslavian Republic of *
1999	Uzbekistan*
2001	Cuba (rejoined)*
2001	Jordan
2001	Philippines
2003	Nigeria
2003	Serbia
2006	Lebanon (rejoining)
2006	Mongolia
2006	Thailand
2009	Costa Rica
2009	Honduras
2009	Panama
2009	Viet Nam
2012	Ethiopia
2012	Kazakhstan
2012	DPR Korea (rejoining)
2015	Colombia (rejoining)
2018	Algeria (rejoining)
2018	Cyprus
2018	Ghana

2018   Jordan (rejoining)
2018   Madagascar
2018   Morocco (rejoining)
2018   Mozambique
2018   Slovenia
2018   Syria
2018   United Arab Emirates

Appendix D: International Schools for Young Astronomers

Nr	Year	Country	Place	Participants	Females	Foreigners	Nationalities
41	2018	Colombia	Socorro, Santander	32	16	16	7
40	2018	Egypt	Cairo	29	13	11	8
39	2017	Ethiopia	Addis Ababa	31	9	14	8
38	2016	Iran	Teheran	34	17	11	7
37	2015	Honduras	Tegucigalpa	30	12		12
36	2014	Thailand	Chiang Mai	41	20	32	14
35	2013	Indonesia	Bandung	39	16	19	11
34	2012	South Africa	Cape Town/ Sutherland	31	4	18	7
33	2011	China	Lijang	37	9	19	10
32	2010	Armenia	Byurakan	48	27	15	21
31	2009	Trinidad	University of the West Indies	31	16	22	12
30	2008	Turkey	Istanbul (Yis)	34	13	18	12
29	2007	Malaysia	Kuala Lumpur/ Langkawi	38	9	28	12
28	2005	Mexico	INAOE	46	17	20	10
27	2004	Morocco	Al Akhawayn Univ-	29	9	18	13
26	2002	Argentina	Casleo	28	10	14	8
25	2001	Thailand	Chiang Mai	36	19	17	9
24	1999	Romania	Buchariste	40	22	18	8

(continued)

© Springer Nature Switzerland AG 2019
J. Andersen et al., *The International Astronomical Union*,
https://doi.org/10.1007/978-3-319-96965-7

Nr	Year	Country	Place	Participants	Females	Foreigners	Nationalities
23	1997	Iran	Zanjan	38	12	14	8
22	1995	Brazil	Belo Horizonte/ Serra Piedade	38	15	20	15
21	1994	Egypt	Cairo/ Kottamia Observatory	41	10	12	13
20	1994	India	Pune	35	11	25	13
19	1992	China	Beijing/ Xinglong Observatory	30	9	17	12
18	1990	Morocco	Marrakech	53	7	39	14
17	1990	Malaysia	Kuala Lumpur/ Melaka	27	6	12	8
16	1989	Cuba	Havana	55	23	23	6
15	1986	Portugal	Espinho	30	19	20	7
14	1986	China	Beijing	52	–	10	6
13	1983	Indonesia	Lembang	21	–	–	5
12	1981	Egypt	Cairo	28	3	9	9
11	1980	Yugoslavia	Hvar	46	13	35	13
10	1979	Spain	Tenerife	63	20	12	6
9	1978	Nigeria	Nsukka	28	2	3	4
8	1977	Brazil	Rio de Janeiro	29	9	14	7
7	1975	Greece	Athens/Thera	74	21	35	16
6	1974	Argentina	San Miguel	60	11	21	7
5	1973	Indonesia	Lembang	8	3	3	4
4	1970	Argentina	Cordoba	23	3	18	5
3	1969	India	Hyderabad	23	2	5	5
2	1968	Italy	Arcetri	10	–	10	7
1	1967	U.K.	Manchester	12	–	12	8

See https://www.iau.org/science/grants_prizes/iau_grants/international_school/list/

Bibliography

The IAU Archives are located at the IAU office in the *Institut Astrophysique de Paris*. The 'old part' refers to the papers that were collected and described by Adriaan Blaauw. Since 1998, they were stored by the *Académie des Sciences*, but they have now been returned to the IAU office. The new part of the archive has been collected and described by IAU archivist Ginette Rude.

AIP interview refers to Interviews from the Oral History Project of the American Institute of Physics, College Park, MD. Transcriptions of the interviews can be accessed via: http://www.aip.org/history/ohilist/transcripts.html

GA Newspapers can be accessed via: https://www.iau.org/publications/iau/ga_newspapers/. The page numbers refer to the page number of these scanned versions.

IB: IAU Information Bulletins. Nos. 75-112 can be accessed via: https://www.iau.org/publications/iau/information_bulletins/

Proceedings of the General Assemblies can be accessed via: https://www.iau.org/publications/iau/transactions_b/

S349 refers to presentations at the IAU Symposium 349 *Under One Sky* in Vienna in 2018, of which the Transactions are forthcoming (2019).

References

Agar, J.: Science and Spectacle: The Work of Jodrell Bank in Post-war British Culture. Harwood Academic Publishers, Amsterdam (1998)

Baneke, D.: Teach and travel: Leiden Observatory and the renaissance of Dutch astronomy in the interwar years. J. Hist. Astron. **41**, 167–198 (2010)

© Springer Nature Switzerland AG 2019   **349**
J. Andersen et al., *The International Astronomical Union*,
https://doi.org/10.1007/978-3-319-96965-7

Baneke, D.: Space for ambitions: the Dutch Space Program in changing European and Transatlantic contexts. Minerva. **52**(1), 119–140 (2014)

Blaauw, A.: ESO's Early History: The European Southern Observatory from Concept to Reality. European Southern Observatory, Garching bei München (1991)

Blaauw, A.: History of the IAU: The Birth and First Half-Century of the International Astronomical Union. Springer, Dordrecht (1994)

Bobrowsky, P., Rickman, H.: Comet/Asteroid Impacts and Human Society. Springer, Heidelberg (2007)

Brown, M.: How I Killed Pluto and Why It Had It Coming. Spiegel & Grau Trade Paperbacks, New York (2010)

Cesarsky, C.: Women in astronomy: IAU statistics. IAU IB. **106**, 28–34 (2010)

Débarbat, S.: Statistics on women in the IAU membership. In: Heck, A. (ed.) Organizations and Strategies in Astronomy, vol. 5, pp. 189–195. Springer, Dordrecht (2004)

DeGrasse Tyson, N.: The Pluto Files: The Rise and Fall of America's Favorite Planet. W. W. Norton, New York (2009)

DeVorkin, D.H.: Community and spectral classification in astrophysics: the acceptance of E. C. Pickering's system in 1910. Isis. **72**, 29–49 (1981)

DeVorkin, D.H.: Science with a Vengeance: How the Military Created the US Space Sciences After World War II. Springer, New York (1992)

DeVorkin, D.H. (ed.): The American Astronomical Society's First Century. American Institute of Physics, Washington, DC (1999)

DeVorkin, D.H.: Who speaks for astronomy? How astronomers responded to government funding after World War II. Hist. Stud. Phys. Biol. Sci. **31**(1), 52–92 (2000)

Dick, S.J. (ed.): National observatories issue. J. Hist. Astron. **22**(1), 1–4 (1991)

Dick, S.J.: Discovery and Classification in Astronomy: Controversy and Consensus. Cambridge University Press, Cambridge (2013)

Gerbaldi, M.: International schools for young astronomers (ISYA): a programme of the International Astronomical Union 2007. In: Hearnshaw, J.B., Martinez, P. (eds.) IAU Special Session 5, Astronomy for the Developing World, pp. 221–228. International Astronomical Union, Cambridge (2007)

Gerbaldi, M., DeGreve, J.-P., Guinan, E.: International schools for young astronomers, teaching for astronomy development: two programmes of the International Astronomical Union. In: Valls-Gabaud, D., Boksenberg, A. (eds.) The Role of Astronomy in Society and Culture, pp. 642–649. Cambridge University Press, Cambridge (2011)

Gingerich, O.: Brian Marsden 1937–2010: the walking encyclopedia of comets. Nature. **468**, 1042 (2010)

Greenaway, F.: Science International: A History of the International Council of Scientific Unions. Cambridge University Press, Cambridge (1996)

Harwit, M.: Cosmic Discovery: The Search, Scope, and Heritage of Astronomy. Cambridge University Press, Cambridge (1981)

Heck, A.: Organizations and Strategies in Astronomy, vol. I. Springer, Dordrecht (2000)

Hufbauer, K.: Exploring the Sun: Solar Science Since Galileo. Johns Hopkins University Press, Baltimore (1991)

Kragh, H., Smith, R.W.: Who discovered the expanding universe? Hist. Sci. **41**, 141–162 (2003)

Krige, J., Russo, A.: A History of the European Space Agency 1958–1987. ESA Publications, Noordwijk (2000)

Lamy, J.: Adjusting astronomical practices: the "Carte du Ciel", the democratic rules and the circulation of opinions at the end 19th century. In: Valls-Gabaud, D., Boksenberg, A. (eds.) The Role of Astronomy in Society and Culture, pp. 195–201. Cambridge University Press, Cambridge (2011)

Lankford, J.: American Astronomy: Community, Careers and Power, 1859–1940. University of Chicago Press, Chicago (1997)

Letho, O.: Mathematics Without Borders: A History of the International Mathematical Union. Springer, New York (1998)

Logsdon, J.: The development of international cooperation. In: Exploring the Unknown, vol. II, pp. 65–76. Express Publishing, Washington, DC (1996)

Madsen, C.: The Jewel on the Mountaintop: The European Southern Observatory Through Fifty Years. Wiley, Weinheim (2012)

McCarthy, D.D., Kenneth Seidelman, P.: Time: From Earth Rotation to Atomic Physics. Wiley, Hoboken (2009)

McCray, W.P.: Giant Telescopes: Astronomical Ambition and the Promise of Technology. Harvard University Press, Cambridge, MA (2004)

McCray, W.P.: How astronomers digitized the sky. Technol. Cult. **55**(4), 908–944 (2014)

Munns, D.P.D.: A Single Sky: How an International Community Forged the Science of Radio Astronomy. MIT Press, Cambridge, MA (2012)

Paul, E.R.: The Milky Way Galaxy and Statistical Cosmology 1890–1924. Cambridge University Press, Cambridge (1993)

Pfau, W.: The Astronomische Gesellschaft: pieces from its history. In: Heck, A. (ed.) Organizations and Strategies in Astronomy, vol. I, pp. 67–75. Kluwer Academic Publishers, Dordrecht (2000)

Rothenberg, M., Williams, T.R.: Amateurs and the society during the formative years. In: DeVorkin, D.H. (ed.) The American Astronomical Society's First Century, pp. 40–52. American Institute of Physics, Washington, DC (1999)

Russo, P., Christensen, L. L.: International Year of Astronomy 2009: Final Report (2010)

Schroeder-Gudehus, B.: Probing the master narrative of scientific internationalism: nationals and neutrals in the 1920s. In: Lettevall, R., Somsen, G., Widmalm, S. (eds.) Neutrality in Twentieth-Century Europe. Intersections of Science, Culture, and Politics after the First World War, pp. 19–42. Routledge, London (2012)

Smith, R.W.: The Space Telescope: A Study of NASA, Science, Technology, and Politics. Cambridge University Press, Cambridge (1989)

Smith, R.W.: Engines of discovery: scientific instruments and the history of astronomy and planetary science in the united states in the twentieth century. J. Hist. Astron. **28**, 49–77 (1997)

Smith, R.W.: Beyond the big galaxy: the structure of the stellar system. J. Hist. Astron. **37**, 307–342 (2006)

Smith, R.W.: The making of space astronomy: a gift of the cold war. In: Morrison-Low, A.D., Dupré, S., Johnston, S., Strano, G. (eds.) Earth-Bound to Satellite: Telescopes, Skills and Networks, pp. 235–249. Brill Publishers, Leiden (2011)

Sullivan III, W.T.: Cosmic Noise: A History of Early Radio Astronomy. Cambridge University Press, Cambridge (2009)

Thomas, J.H.: The solar physics division. In: DeVorkin, D.H. (ed.) The American Astronomical Society's First Century, pp. 238–251. American Institute of Physics, Washington, DC (1999)

van der Kruit, P.: Jacobus Cornelius Kapteyn: Born Investigator of the Heavens. Springer, Dordrecht (2014)

Matter Index

© Springer Nature Switzerland AG 2019
J. Andersen et al., *The International Astronomical Union*,
https://doi.org/10.1007/978-3-319-96965-7

Author Index

© Springer Nature Switzerland AG 2019
J. Andersen et al., *The International Astronomical Union*,
https://doi.org/10.1007/978-3-319-96965-7